Springer Monographs in Mathematics

T0234327

Vilmos Komornik   Paola Loreti

Fourier Series
in Control Theory

 Springer

Vilmos Komornik
Institut de Recherche Mathématique
    Avancée
Université Louis Pasteur et CNRS
7, rue René·Descartes
67084 Strasbourg Cedex
France

Paola Loreti
Dipartimento di Metodi e Modelli
Matematici per le Scienze Applicate
Universitá degli Studi di Roma "La Sapienza"
Via A. Scarpa, 16
00161 Roma
Italy

ISBN 978-1-4419-1975-5   e-ISBN 978-0-387-27408-9

Mathematics Subject Classification (2000): 49-xx, 93-xx

Komornik, V.
    Fourier series in control theory / Vilmos Komornik and Paola Loreti.
      p. cm. — (Springer monographs in mathematics)
    Includes bibliographical references and index.

    1. Control theory.   2. Fourier series.   I. Loreti, Paola.   II. Title.   III. Series.
    QA402.3.K5785 2004
    003'.5—dc22   2004056525

Printed on acid-free paper.

9 8 7 6 5 4 3 2 1

Springer is a part of Springer Science+Business Media
springeronline.com

Preface

Before showing the Table of contents of this book, we would like briefly to explain our motivation to the subject. In the mid-eighties, J.-L. Lions contributed to the multiplier method by proving powerful and elegant theorems on observability, controllability, and uniform stabilization. His 1988 survey article and monograph also stimulated intensive research activity in the field.

The multiplier method led to great success, but several problems remained unsolved. Starting research on them, we came up with another efficient way to deal with those issues: using a former approach based on harmonic analysis. Indeed, following the influential survey paper of D.L. Russell (1978), many authors had emphasized a classical result of A.E. Ingham (1936) for its simplicity and depth that had proven to be extremely useful in control theory.

In this book, our purpose is to unify, as much as possible, the so-called harmonic (or nonharmonic) analysis method. It is also to make the subject as simple as possible. We start by solving elementary "ad hoc" controllability problems; then we extend the results and the proofs to a general framework. The book contains almost all proofs of the theorems, and only little knowledge of functional analysis is required. Many results presented here are new and still unpublished, while many known results have been rewritten for the purpose of simplification.

The last part of this book is devoted to the exposition and the derivation of some joint results with C. Baiocchi. We would like to take this opportunity to thank him for his precious contribution to our work. We are also grateful to all our students and colleagues for their encouragement as well as for their interest through very useful discussions and comments. Finally, we wish to thank the editorial staff at Springer-Verlag, New York, for their help and support.

Rome and Strasbourg, August 2003

Contents

1

Introduction

Consider the small transversal vibrations of a string with two free endpoints. Denoting by $u(t, x)$ the transversal displacement at time t of the point of abscissa x, it is well known that a suitable linear model is given by the following system:

$$\begin{cases} u_{tt} - u_{xx} = 0 & \text{in} \quad \mathbb{R} \times (0, \ell), \\ u_x(t, 0) = u_x(t, \ell) = 0 & \text{for} \quad t \in \mathbb{R}, \\ u(0, x) = u_0(x) & \text{for} \quad x \in (0, \ell), \\ u_t(0, x) = u_1(x) & \text{for} \quad x \in (0, \ell). \end{cases} \tag{1.1}$$

Here ℓ denotes the length of the string, u_0 and u_1 denote the initial data, and the subscripts t and x stand for time and spatial derivations, respectively. See Figure 1.1 for a possible position of the string.

Assume that we can observe the oscillations of the left endpoint of the string only during some interval of time $0 \leq t \leq T$. Can we identify the unknown initial data? In other words, is the linear map

$$(u_0, u_1) \mapsto u(\cdot, 0)|_{(0,T)} \tag{1.2}$$

one-to-one in suitable "natural" function spaces? And what can we say about the continuity of this map and of its inverse (if it exists)?

The problem can be solved easily by using Fourier series. Indeed, choose $\ell = \pi$ for simplicity of the formulae and introduce the Hilbert·spaces

$$H := \left\{ v \in L^2(0, \pi) \; : \; \int_0^\pi v(x) \, dx = 0 \right\}, \quad \|v\|_H := \left(\int_0^\pi |v(x)|^2 \, dx \right)^{1/2},$$

and

$$V := \left\{ v \in H^1(0, \pi) \; : \; \int_0^\pi v(x) \, dx = 0 \right\}, \quad \|v\|_V := \left(\int_0^\pi |v'(x)|^2 \, dx \right)^{1/2}.$$

Introducing the initial energy of the solution of (1.1) by the formula

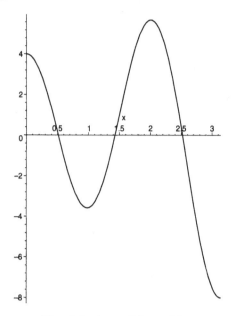

Fig. 1.1. A possible position

$$E_0 := \frac{1}{2}(\|u_0\|_V^2 + \|u_1\|_H^2),$$

we have the following proposition:

Proposition 1.1. *If $T \geq 2\pi$, then the map (1.2) is one-to-one from $V \times H$ into $H^1(0,T)$. Moreover, there exist two constants $c_1, c_2 > 0$ such that the solutions of (1.1) satisfy the estimates*

$$c_0 E_0 \leq \int_0^T |u_t(t,0)|^2 \, dt \leq c_2 E_0$$

for all $(u_0, u_1) \in V \times H$.

Proof. Since the functions

$$\cos kx, \quad k = 1, 2, \ldots,$$

form an orthogonal basis in both H and V, the solution of (1.1) is given by the series

$$u(t,x) = \sum_{k=1}^{\infty}(a_k \cos kt + b_k \sin kt) \cos kx$$

with suitable real coefficients a_k and b_k, depending on the initial data.

Using the orthogonality of the basis functions, we have the equalities

$$\|u_1\|_H^2 = \int_0^\pi |u_t(0,x)|^2 \, dx$$

$$= \int_0^\pi \left| \sum_{k=1}^\infty k b_k \cos kx \right|^2 dx$$

$$= \sum_{k=1}^\infty k^2 b_k^2 \int_0^\pi \cos^2 kx \, dx$$

$$= \frac{\pi}{2} \sum_{k=1}^\infty k^2 b_k^2.$$

Similarly, using the orthogonality of the functions

$$\sin kx, \quad k = 1, 2, \ldots,$$

we obtain that

$$\|u_0\|_V^2 = \int_0^\pi |u_x(0,x)|^2 \, dx$$

$$= \int_0^\pi \left| \sum_{k=1}^\infty -k a_k \sin kx \right|^2 dx$$

$$= \sum_{k=1}^\infty k^2 a_k^2 \int_0^\pi \sin^2 kx \, dx$$

$$= \frac{\pi}{2} \sum_{k=1}^\infty k^2 a_k^2.$$

Hence

$$E_0 = \frac{\pi}{4} \sum_{k=1}^\infty k^2 (a_k^2 + b_k^2). \tag{1.3}$$

Furthermore, for any positive integer M, the functions

$$\cos kt \quad \text{and} \quad \sin kt, \quad k = 1, 2, \ldots,$$

also form an orthogonal system in $L^2(0, 2M\pi)$, so that

$$\int_0^{2M\pi} |u_t(t,0)|^2 \, dt = \int_0^{2M\pi} \left| \sum_{k=1}^\infty k(-a_k \sin kt + b_k \cos kt) \right|^2 dt$$

$$= \sum_{k=1}^\infty k^2 \int_0^{2M\pi} a_k^2 \sin^2 kt + b_k^2 \cos^2 kt \, dt$$

$$= M\pi \sum_{k=1}^\infty k^2 (a_k^2 + b_k^2),$$

i.e.,

$$\int_0^{2M\pi} |u_t(t,0)|^2 \ dt = M\pi \sum_{k=1}^{\infty} k^2(a_k^2 + b_k^2). \tag{1.4}$$

We deduce from (1.3) and (1.4) the identity

$$\int_0^{2M\pi} |u_t(t,0)|^2 \ dt = 4ME_0.$$

Denoting by M the integer part of $T/2\pi$, so that $2M\pi \leq T < 2(M+1)\pi$, and using the nonnegativity of the function under the integral sign, it follows that

$$4ME_0 \leq \int_0^T |u_t(t,0)|^2 \ dt \leq 4(M+1)E_0.$$

Several remarks are in order:

- By analyzing the above proof, one can show that the hypothesis $T \geq 2\pi$ is optimal.
- The above method can be easily adapted to other boundary conditions.
- The method can also be adapted (after some algebraic manipulations) to cases in which we observe both endpoints.
- We can also apply this approach to vibrating *circular or spherical membranes*, and to vibrating bodies occupying a *ball* in their rest position.
- Proposition 1.1 can also be established in at least two other elementary ways: by applying either d'Alembert's formula or the multiplier method. It would be interesting to compare the relative advantages and drawbacks of the three methods.
- It would be more natural to consider initial data in the larger and more natural spaces $H^1(0,\pi)$ and $L^2(0,\pi)$ instead of V and H. However, this leads to some difficulties: observe that the constant functions solve (1.1), but they do not satisfy the first inequality in the proposition.
- Slight changes in the state equation, such as the addition of lower-order terms, also lead to serious technical difficulties. For example by replacing the original state equation by

$$u_{tt} - u_{xx} + u = 0 \quad \text{in} \quad \mathbb{R} \times (0,\pi),$$

the solutions of the modified system are given by the series

$$u(t,x) = \sum_{k=1}^{\infty} (a_k \cos \omega_k t + b_k \sin \omega_k t) \cos kx$$

with

$$\omega_k = \sqrt{k^2 + 1}.$$

Now the functions

$$\cos \omega_k t \quad \text{and} \quad \sin \omega_k t, \quad k = 1, 2, \ldots,$$

are no longer orthogonal in any interval $(0, T)$, so that the integrals

$$\int_0^T |u_t(t, 0)|^2 \, dt$$

cannot be evaluated by a simple application of Parseval's equality.

The above approach can also be adapted to the study of vibrating *beams*. As an illustration, consider the following linear model:

$$\begin{cases} u_{tt} + u_{xxxx} = 0 & \text{in} \quad \mathbb{R} \times (0, \pi), \\ u_x(t, 0) = u_{xxx}(t, 0) = u_x(t, \pi) = u_{xxx}(t, \pi) = 0 & \text{for} \quad t \in \mathbb{R}, \\ u(0, x) = u_0(x) & \text{for} \quad x \in (0, \pi), \\ u_t(0, x) = u_1(x) & \text{for} \quad x \in (0, \pi). \end{cases} \quad (1.5)$$

Now introducing the Hilbert spaces

$$H := \left\{ v \in L^2(0, \pi) \ : \ \int_0^\pi v(x) \, dx = 0 \right\}$$

and

$$V := \left\{ v \in H^2(0, \pi) \ : \ v'(0) = v'(\pi) = \int_0^\pi v(x) \, dx = 0 \right\}$$

with the norms

$$\|v\|_H := \left(\int_0^\pi |v(x)|^2 \, dx \right)^{1/2} \quad \text{and} \quad \|v\|_V := \left(\int_0^\pi |v''(x)|^2 \, dx \right)^{1/2},$$

and the initial energy of the solution of (1.5) by the formula

$$E_0 := \frac{1}{2} \left(\|u_0\|_V^2 + \|u_1\|_H^2 \right),$$

we have the following result:

Proposition 1.2. *If $T \geq 2\pi$, then the map*

$$(u_0, u_1) \mapsto u(\cdot, 0)|_{(0,T)}$$

is one-to-one from $V \times H$ into $H^1(0, T)$. Moreover, there exist two constants $c_1, c_2 > 0$ such that the solutions of (1.5) satisfy the estimates

$$c_1 E_0 \leq \int_0^T |u_t(t, 0)|^2 \, dt \leq c_2 E_0$$

for all $(u_0, u_1) \in V \times H$.

Proof. The solution of (1.5) is given by the series

$$u(t, x) = \sum_{k=1}^{\infty} (a_k \cos k^2 t + b_k \sin k^2 t) \cos kx$$

with suitable real coefficients a_k and b_k, depending on the initial data.

Adapting the computations of the preceding proof, we obtain easily the equalities

$$\|u_1\|_H^2 = \frac{\pi}{2} \sum_{k=1}^{\infty} k^4 b_k^2, \quad \|u_0\|_V^2 = \frac{\pi}{2} \sum_{k=1}^{\infty} k^4 a_k^2,$$

and

$$\int_0^{2M\pi} |u_t(t, 0)|^2 \ dt = M\pi \sum_{k=1}^{\infty} k^2 (a_k^2 + b_k^2).$$

It follows that

$$\int_0^{2M\pi} |u_t(t, 0)|^2 \ dt = 4ME_0$$

for every positive integer M, and we conclude that

$$4ME_0 \leq \int_0^T |u_t(t, 0)|^2 \ dt \leq 4(M+1)E_0$$

if M denotes the integer part of $T/2\pi$.

Again, several remarks can be made:

- As we will see later, this time the hypothesis $T \geq 2\pi$ is *not* optimal.
- The method can again be adapted to other boundary conditions.
- We can also apply this approach to vibrating *circular or spherical plates*.
- Proposition 1.2 can also be established by applying the multiplier method. We shall discuss later the relative advantages and drawbacks of these two methods.
- It would be more natural to consider initial data in larger spaces instead of V and H, by removing the conditions

$$\int_0^\pi u_0(x) \ dx = \int_0^\pi u_1(x) \ dx = 0.$$

However, this leads to interesting technical difficulties.
- The presence of lower-order terms in the state equation leads again to serious technical difficulties.

The purpose of this book is to address the above remarks by generalizing the above simple approach based on Parseval's equality. Relatively simple tools will already enable us to obtain much more general and precise results. Further generalizations will allow us to solve various controllability problems concerning vibrating strings, beams, membranes, plates, shells, or systems of them. For many models studied in this volume, the otherwise very powerful *multiplier method*[1] does not seem to apply.

The applications of the methods developed here are not limited to control theory. As an example, we shall give a new simple proof of a celebrated generalization of Bernstein of Pólya's theorem on the singularities of Dirichlet series.

We assume that the reader is familiar with the basic results on linear partial differential equations, and with the simplest Lebesgue and Sobolev spaces such as

$$L^2(\Omega), \quad H^1(\Omega), \quad H^2(\Omega), \quad H_0^1(\Omega), \quad H^2(\Omega) \cap H_0^1(\Omega),$$

and the dual space

$$H^{-1}(\Omega) := (H_0^1(\Omega))',$$

where Ω is a nonempty bounded open domain of $\mathbb{R}^N$ having a sufficiently smooth boundary Γ as exposed, e.g., in [31] and [98].

For the convenience of the reader, we give a short review of some parts of linear control theory in Chapter 2. In particular, we present briefly the main ideas of the Hilbert Uniqueness Method of Lions, which reduces many problems of controllability to the observability of dual systems, and of an analogous method developed recently, which does the same for stabilizability problems. This enables us to concentrate on the observability problems in the rest of the book: using the general theory, the reader can readily deduce from them the corresponding controllability and stabilization results.

---

[1]We refer to [96], [97], or [67] for an introduction to the multiplier method. See also Sections 6.6 and 6.7 of this book, pp. 114 and 118.

2

Observation, Control, and Stabilization

The aim of this chapter is to review some general results of control theory concerning the relations among the three fundamental concepts in the title. Since in this book we consider only evolutionary problems with time-reversible dynamics, we restrict ourselves to this framework. We present briefly the duality between the notions of observability and controllability, which lies at the basis of the celebrated Hilbert uniqueness method of J.-L. Lions. Then we also explain the main ideas of an analogous method developed more recently in the framework of distributed parameter systems, which reduces many problems of stabilization to problems of observability. These two main principles allow us to concentrate in the rest of the book exclusively on questions of observability.

Since the results of this chapter will not be used in the sequel, some proofs are omitted. The interested reader may find them in the works [96], [97] of Lions or in the textbook [67] (concerning controllability) and in the papers [70] and [100] (concerning stabilizability).

2.1 Well-Posedness of Linear Evolutionary Equations

We are going to investigate the well-posedness of the problem[1]

$$U' = \mathcal{A}U, \quad U(0) = U_0, \tag{2.1}$$

in a *complex* Hilbert space $\mathcal{H}$, where $\mathcal{A}$ is a (bounded or unbounded) linear operator defined on some linear subspace of $\mathcal{H}$, with values in $\mathcal{H}$. Let us make the following asumption[2]:

---

[1] We shall often use the notation U' for the time derivative dU/dt.

[2] We use semigroups only in this chapter, but the results presented here will not be used in the sequel. The rest of the book can be read without any knowledge of the theory of semigroups.

(H1) The operator $\mathcal{A}$ generates a *strongly continuous group* of automorphisms $e^{t\mathcal{A}}$ in $\mathcal{H}$.

Examples.

- If $\mathcal{H}$ is finite-dimensional, then every linear operator in $\mathcal{H}$ satisfies hypothesis (H1).
- Hypothesis (H1) is also satisfied if $\mathcal{A}$ is a *skew-adjoint*[3] operator having a compact resolvent.

Under the assumption (H1) the problem (2.1) is well-posed in the following sense:

Theorem 2.1. *Assume (H1). Given $U_0 \in \mathcal{H}$ arbitrarily, the problem (2.1) has a unique continuous solution $U : \mathbb{R} \to \mathcal{H}$, satisfying the estimate*

$$\|U(t)\| \le Me^{\alpha|t|}\|U_0\|$$

with suitable constants $M \ge 1$ and $\alpha \ge 0$, independent of the particular choice of the initial data $U_0 \in \mathcal{H}$.
  If $U_0 \in D(\mathcal{A})$, then the solution is also continuously differentiable.
  If $\mathcal{A}$ is skew-adjoint, then we even have $\|U(t)\| = \|U_0\|$ for all $t \in \mathbb{R}$.

Proof. See any textbook on semigroup theory, e.g., Pazy [112].

Remark. Let us also recall that more generally, the inhomogeneous problem

$$U' = \mathcal{A}U + F, \quad U(0) = U_0,$$

also has a unique continuous solution $U : \mathbb{R} \to \mathcal{H}$ for any given $U_0 \in \mathcal{H}$ and a locally integrable function $F : \mathbb{R} \to \mathcal{H}$, given by the formula of *variation of constants*:

$$U(t) = e^{t\mathcal{A}}U_0 + \int_0^t e^{(t-s)\mathcal{A}}F(s)\,ds, \quad t \in \mathbb{R}.$$

Furthermore, if $U_0 \in D(\mathcal{A})$ and $F : \mathbb{R} \to \mathcal{H}$ is continuously differentiable, then the solution is also continuously differentiable; see Pazy [112], Corollary 2.5, p. 107.

Let us give two important examples.

2.1.1 Wave Equation with Homogeneous Dirichlet Boundary Conditions

Consider the problem

---

[3]We recall that an operator $\mathcal{A}$ is skew-adjoint if $i\mathcal{A}$ is autoadjoint.

$$\begin{cases} u'' - \Delta u = 0 & \text{in } \mathbb{R} \times \Omega, \\ u = 0 & \text{on } \mathbb{R} \times \Gamma, \\ u(0) = u_0 \quad \text{and} \quad u'(0) = u_1 & \text{in } \Omega, \end{cases} \qquad (2.2)$$

where Ω is a nonempty bounded open set in $\mathbb{R}^N$ with boundary Γ. In order to simplify the exposition we usually assume that Ω is of class C^∞. Here and in the sequel, we use the notation

$$u' := \frac{\partial u}{\partial t}, \quad u'' := \frac{\partial^2 u}{\partial t^2}, \quad \text{and} \quad \Delta u := \sum_{j=1}^{N} \frac{\partial^2 u}{\partial x_j^2}$$

for the time derivatives and for the *Laplacian* of u. The *energy* of the solution is defined by

$$E(t) := \frac{1}{2} \int_\Omega |\nabla u(t,x)|^2 + |u'(t,x)|^2 \; dx, \quad t \in \mathbb{R}.$$

Proposition 2.2. *If*

$$u_0 \in H_0^1(\Omega) \quad \text{and} \quad u_1 \in L^2(\Omega),$$

then (2.2) has a unique solution satisfying

$$u \in C(\mathbb{R}; H_0^1(\Omega)) \cap C^1(\mathbb{R}; L^2(\Omega)).$$

Moreover, the energy of the solution is conserved: $E(t) = E(0)$ for all $t \in \mathbb{R}$.

Proof. We rewrite the problem (2.2) in the form (2.1) by setting

$$U := (u, u'), \quad U_0 := (u_0, u_1), \quad \text{and} \quad \mathcal{A}(u,v) := (v, \Delta u).$$

One may readily verify that $\mathcal{A}$ becomes a skew-adjoint operator with a compact resolvent in the Hilbert space

$$\mathcal{H} := H_0^1(\Omega) \times L^2(\Omega), \quad \|(u,v)\|_{\mathcal{H}} := \left(\int_\Omega |\nabla u|^2 + |v|^2 \; dx \right)^{1/2},$$

if we define its domain by

$$D(\mathcal{A}) := \{(u,v) \in \mathcal{H} \ : \ \mathcal{A}(u,v) \in \mathcal{H}\} = \left(H^2(\Omega) \cap H_0^1(\Omega) \right) \times H_0^1(\Omega).$$

Hence we can apply Theorem 2.1. The energy conservation follows from the equality

$$E(t) = \frac{1}{2} \|U(t)\|^2.$$

2.1.2 A Petrovsky System with Hinged Boundary Conditions

Now consider the problem

$$\begin{cases} u'' + \Delta^2 u = 0 & \text{in} \quad \mathbb{R} \times \Omega, \\ u = \Delta u = 0 & \text{on} \quad \mathbb{R} \times \Gamma, \\ u(0) = u_0 \quad \text{and} \quad u'(0) = u_1 & \text{in} \quad \Omega. \end{cases} \tag{2.3}$$

Define the energy of the solutions by the formula

$$E(t) := \frac{1}{2} \int_{\Omega} |u(t,x)|^2 + |\Delta u(t,x)|^2 + |\nabla u'(t,x)|^2 \, dx, \quad t \in \mathbb{R}.$$

Proposition 2.3. *If*

$$u_0 \in H_0^1(\Omega) \quad \text{and} \quad u_1 \in H^{-1}(\Omega),$$

then (2.3) has a unique solution

$$u \in C(\mathbb{R}; H_0^1(\Omega)) \cap C^1(\mathbb{R}; H^{-1}(\Omega)).$$

Moreover, the energy of the solution is conserved: $E(t) = E(0)$ *for all* $t \in \mathbb{R}$.

Proof. We rewrite the problem (2.3) in the form (2.1) by setting

$$U := (u, u'), \quad U_0 := (u_0, u_1), \quad \text{and} \quad \mathcal{A}(u,v) := (v, -\Delta^2 u).$$

One may readily verify that $\mathcal{A}$ becomes a skew-adjoint operator with a compact resolvent in the Hilbert space

$$\mathcal{H} := H_0^1(\Omega) \times H^{-1}(\Omega), \quad \|(u,v)\|_{\mathcal{H}} := \left(\int_{\Omega} |\nabla u|^2 + |\nabla \Delta^{-1} v|^2 \, dx \right)^{1/2},$$

where Δ^{-1} denotes[4] the inverse of the restriction of Δ to $H_0^1(\Omega)$, if we define its domain by

$$D(\mathcal{A}) := \{(u,v) \in \mathcal{H} \ : \ \mathcal{A}(u,v) \in \mathcal{H}\}$$
$$= \{u \in H^3(\Omega) \ : \ u = \Delta u = 0 \quad \text{on} \quad \Gamma\} \times H_0^1(\Omega).$$

Since

$$E(t) = \frac{1}{2} \|U(t)\|^2$$

by definition, we conclude the proof by applying Theorem 2.1.

---

[4]We recall that Δ is a linear ismometry of $H_0^1(\Omega)$ onto its dual space $H^{-1}(\Omega)$.

2.2 Weak Solutions of Dual Problems

In order to obtain satisfactory controllability and stabilizability results, we often have to solve *inhomogeneous* problems with *irregular* data. We define solutions to such problems by applying the method of *transposition* or *duality*.

Consider again a first-order equation

$$U' = \mathcal{A}U, \quad U(0) = U_0, \tag{2.4}$$

satisfying assumption (H1) of Theorem 2.1. Furthermore, let $\mathcal{B}$ be another linear operator, defined on some linear subspace $D(\mathcal{B})$ of $\mathcal{H}$ with values in another Hilbert space $\mathcal{G}$, satisfying the following hypotheses:

(H2) $D(\mathcal{A}) \subset D(\mathcal{B})$, and there exists a constant c such that $\|\mathcal{B}U_0\|_{\mathcal{G}} \leq c\|\mathcal{A}U_0\|_{\mathcal{H}}$ for all $U_0 \in D(\mathcal{A})$.

(H3) There exist an interval[5] I and a constant c_I such that the solutions of (2.4) satisfy the inequality

$$\|\mathcal{B}U\|_{L^2(I;\mathcal{G})} \leq c_I \|U_0\|_{\mathcal{H}}$$

for all $U_0 \in D(\mathcal{A})$.

The operator $\mathcal{B}$ is usually called an *observability* operator: we may think that we can observe only $\mathcal{B}U$ and not the whole solution U.

Remarks.

- Hypothesis (H2) ensures that hypothesis (H3) is meaningful.
- Hypothesis (H3) is an abstract form of the *direct* inequalities in the terminology of Lions [96], [97]. It is also called an *admissibility* inequality because it allows us to define $\mathcal{B}U$ as an element of $L^2(I;\mathcal{G})$ for all $U_0 \in \mathcal{H}$, by a density argument.
- The particular interval I does not play an important role in hypothesis (H3). Indeed, since the operator $\mathcal{A}$ does not depend on time, if hypothesis (H3) is satisfied for *some* I, then it is also satisfied for *every* interval J, of arbitrary length, with another constant c_J depending on J. Indeed, let us cover J with a finite number of translates $I + t_1, \ldots, I + t_n$ of I. Since $U_j(t) := U(t+t_j)$ solves (2.4) with the initial data $U(t_j)$ instead of U_0, we have

---

[5]Throughout this book all intervals are assumed to be bounded and nondegenerate, i.e., having a finite positive length.

$$\|\mathcal{B}U\|_{L^2(I_1;\mathcal{G})} \leq \sum_{j=1}^{n}\|\mathcal{B}U\|_{L^2(I+t_j;\mathcal{G})}$$

$$= \sum_{j=1}^{n}\|\mathcal{B}U_j\|_{L^2(I;\mathcal{G})}$$

$$\leq c\sum_{j=1}^{n}\|U(t_j)\|_{\mathcal{H}}$$

$$\leq c_J\|U_0\|_{\mathcal{H}},$$

where the constant c_J also depends on the numbers t_j. Consequently, hypothesis (H3) allows us to define $\mathcal{B}U$ for all $U_0 \in \mathcal{H}$ as an element of $L^2_{\mathrm{loc}}(\mathbb{R};\mathcal{G})$.

Under the assumptions (H1)–(H3), we are going to define the solution of the *dual* problem

$$V' = -\mathcal{A}^*V + \mathcal{B}^*W, \quad V(0) = V_0. \tag{2.5}$$

for every $V_0 \in \mathcal{H}'$ and $W \in L^2_{\mathrm{loc}}(\mathbb{R};\mathcal{G}')$. Here $\mathcal{H}'$, $\mathcal{G}'$ denote the dual spaces of $\mathcal{H}$, $\mathcal{G}$, and $\mathcal{A}^*$, $\mathcal{B}^*$ denote the adjoints of $\mathcal{A}$ and $\mathcal{B}$.

The operator $\mathcal{B}^*$ is usually called a *controllability* operator: we may think that we can act on the system by choosing a *control* W. Proceeding formally, if U solves (2.4) and V solves (2.5), then for each $S \in \mathbb{R}$ we have the identity

$$\langle V(S), U(S)\rangle_{\mathcal{H}',\mathcal{H}} = \langle V_0, U_0\rangle_{\mathcal{H}',\mathcal{H}} + \int_0^S \langle W(t), \mathcal{B}U(t)\rangle_{\mathcal{G}',\mathcal{G}}\, dt. \tag{2.6}$$

Indeed, we have

$$0 = \int_0^S \langle V(t), U'(t) - \mathcal{A}U(t)\rangle_{\mathcal{H}',\mathcal{H}}\, dt$$

$$= [\langle V(t), U(t)\rangle_{\mathcal{H}',\mathcal{H}}]_0^S - \int_0^S \langle V'(t), U(t)\rangle_{\mathcal{H}',\mathcal{H}} + \langle V(t), \mathcal{A}U(t)\rangle_{\mathcal{H}',\mathcal{H}}\, dt$$

$$= [\langle V(t), U(t)\rangle_{\mathcal{H}',\mathcal{H}}]_0^S - \int_0^S \langle V'(t) + \mathcal{A}^*V(t), U(t)\rangle_{\mathcal{H}',\mathcal{H}}\, dt$$

$$= [\langle V(t), U(t)\rangle_{\mathcal{H}',\mathcal{H}}]_0^S - \int_0^S \langle \mathcal{B}^*W(t), U(t)\rangle_{\mathcal{H}',\mathcal{H}}\, dt$$

$$= [\langle V(t), U(t)\rangle_{\mathcal{H}',\mathcal{H}}]_0^S - \int_0^S \langle W(t), \mathcal{B}U(t)\rangle_{\mathcal{G}',\mathcal{G}}\, dt.$$

Hence we *define* a solution of (2.5) as a function $V : \mathbb{R} \to \mathcal{H}'$ satisfying the identity (2.6) for all $U_0 \in \mathcal{H}$ and for all $S \in \mathbb{R}$. This definition is justified by the following theorem:

Theorem 2.4. *Assume (H1)–(H3). If $V_0 \in \mathcal{H}'$ and $W \in L^2_{\mathrm{loc}}(\mathbb{R};\mathcal{G}')$, then the problem (2.5) has a unique solution. Moreover, the function $V : \mathbb{R} \to \mathcal{H}'$ is continuous.*

Proof. As a consequence of hypothesis (H3), the right-hand side of (2.6) defines a bounded linear form of $U_0 \in \mathcal{H}$ for each fixed real S. Since the linear map e^{SA} is an automorphism of $\mathcal{H}$ by Theorem 2.1, the right-hand side of (2.6) is also a bounded linear form of $U(S) \in \mathcal{H}$. Denoting this form by $V(S) \in \mathcal{H}'$, we conclude that (2.6) is satisfied.

Since the right-hand side of (2.6) depends continuously on S, the function $V : \mathbb{R} \to \mathcal{H}'$ is continuous.

Let us turn to the examples.

2.2.1 Wave Equation with Inhomogeneous Dirichlet Boundary Conditions

Consider the following two problems:

$$\begin{cases} v'' - \Delta v = 0 & \text{in} \quad \mathbb{R} \times \Omega, \\ v = w & \text{on} \quad \mathbb{R} \times \Gamma, \\ v(0) = v_0 \quad \text{and} \quad v'(0) = v_1 & \text{in} \quad \Omega; \end{cases} \tag{2.7}$$

and

$$\begin{cases} u'' - \Delta u = 0 & \text{in} \quad \mathbb{R} \times \Omega, \\ u = 0 & \text{on} \quad \mathbb{R} \times \Gamma, \\ u(0) = u_0 \quad \text{and} \quad u'(0) = u_1 & \text{in} \quad \Omega. \end{cases} \tag{2.8}$$

If v solves (2.7) and u solves (2.8), then we can make the following formal computation for every $S \in \mathbb{R}$:

$$\begin{aligned} 0 &= \int_0^S \int_\Omega (v'' - \Delta v) u \; dx \; dt \\ &= \left[\int_\Omega v'u - vu' \; dx \right]_0^S + \int_0^S \int_\Omega v(u'' - \Delta u) \; dx \; dt \\ &\quad + \int_0^S \int_\Gamma -(\partial_\nu v)u + v(\partial_\nu u) \; d\Gamma \; dt. \end{aligned}$$

Using the initial and boundary conditions for v and for u, we therefore conclude that

$$\int_\Omega -v'(S)u(S) + v(S)u'(S) \; dx$$

$$= \int_\Omega -v_1 u_0 + v_0 u_1 \; dx + \int_0^S \int_\Gamma w\partial_\nu u \; d\Gamma \; dt. \tag{2.9}$$

Using the notation

$$U := (u, u'), \quad U_0 := (u_0, u_1), \quad \mathcal{H} := H_0^1(\Omega) \times L^2(\Omega),$$

and

$$\mathcal{A}(u, v) := (v, \Delta u), \quad D(\mathcal{A}) := \left(H^2(\Omega) \cap H_0^1(\Omega) \right) \times H_0^1(\Omega)$$

of Section 2.1.1 (p. 10), defining

$$D(\mathcal{B}) := D(\mathcal{A}), \quad \mathcal{B}(u, v) := \frac{\partial u}{\partial \nu}, \quad \mathcal{G} := L^2(\Gamma),$$

and putting

$$V := (-v', v), \quad V_0 := (-v_1, v_0), \quad \text{and} \quad W := w,$$

we see that (2.7), (2.8), and (2.9) take the form of (2.4), (2.5), and (2.6). Therefore, it is natural to interpret the problem (2.7) as the dual (2.8).

The assumptions (H1)–(H3) are satisfied. Indeed, (H1) was already verified in Section 2.1.1, while (H2) follows from the usual trace theorem. Finally, (H3) is equivalent to the following so-called *hidden regularity* theorem, due to Lasiecka and Triggiani [90]:[6]

Theorem 2.5. *The solutions of* (2.8) *satisfy the* direct *inequality*

$$\int_I \int_\Gamma \left| \frac{\partial u}{\partial \nu} \right|^2 \, d\Gamma \, dt \leq c \int_\Omega |\nabla u_0|^2 + |u_1|^2 \, dx$$

for every time interval[7] I, with a constant c depending only on the length $|I|$ of I.

Applying Theorem 2.4 we obtain the following result:

Proposition 2.6. *For arbitrary*

$$v_0 \in L^2(\Omega)', \quad v_1 \in H^{-1}(\Omega), \quad \text{and} \quad w \in L^2_{\text{loc}}(\mathbb{R}; L^2(\Gamma)'),$$

the problem (2.7) *has a unique solution satisfying*

$$y \in C(\mathbb{R}; L^2(\Omega)') \cap C^1(\mathbb{R}; H^{-1}(\Omega)).$$

2.2.2 A Petrovsky System with Inhomogeneous Boundary Conditions

Here we consider the problems

---

[6]A simpler proof, using multipliers, was subsequently given by Lions [94]. See also [96], [97], or [67], p. 20.

[7]We recall that in this book all intervals are assumed to be bounded and nondegenerate.

$$\begin{cases} v'' + \Delta^2 v = 0 & \text{in } \mathbb{R} \times \Omega, \\ v = 0 \quad \text{and} \quad \Delta v = w & \text{on } \mathbb{R} \times \Gamma, \\ v(0) = v_0 \quad \text{and} \quad v'(0) = v_1 & \text{in } \Omega, \end{cases} \tag{2.10}$$

and

$$\begin{cases} u'' + \Delta^2 u = 0 & \text{in } \mathbb{R} \times \Omega, \\ u = \Delta u = 0 & \text{on } \mathbb{R} \times \Gamma, \\ u(0) = u_0 \quad \text{and} \quad u'(0) = u_1 & \text{in } \Omega. \end{cases} \tag{2.11}$$

If v solves (2.10) and u solves (2.11), then we can make the following formal computation for every $S \in \mathbb{R}$:

$$\begin{aligned} 0 &= \int_0^S \int_\Omega (v'' + \Delta^2 v) u \; dx \; dt \\ &= \left[\int_\Omega v'u - vu' \; dx \right]_0^S + \int_0^S \int_\Omega v(u'' + \Delta^2 u) \; dx \; dt \\ &\quad + \int_0^S \int_\Gamma (\partial_\nu \Delta v) u - (\Delta v)(\partial_\nu u) + (\partial_\nu v)(\Delta u) - v(\partial_\nu \Delta u) \; d\Gamma \; dt. \end{aligned}$$

Using the initial and boundary conditions for v and for u, we conclude that

$$\begin{aligned} \int_\Omega -v'(S)u(S) &+ v(S)u'(S) \; dx \\ &= \int_\Omega -v_1 u_0 + v_0 u_1 \; dx - \int_0^S \int_\Gamma w \partial_\nu u \; d\Gamma \; dt. \end{aligned} \tag{2.12}$$

Using the notation

$$U := (u, u'), \quad U_0 := (u_0, u_1), \quad \mathcal{H} := H_0^1(\Omega) \times H^{-1}(\Omega),$$

and

$$\mathcal{A}(u, v) := (v, -\Delta^2 u),$$
$$D(\mathcal{A}) := \{ u \in H^3(\Omega) \; : \; u = \Delta u = 0 \quad \text{on} \quad \Gamma \} \times H_0^1(\Omega)$$

of Section 2.1.2 (p. 12), defining

$$D(\mathcal{B}) := D(\mathcal{A}), \quad \mathcal{B}(u, v) := -\frac{\partial u}{\partial \nu}, \quad \mathcal{G} := L^2(\Gamma),$$

and putting

$$V := (-v', v), \quad V_0 := (-v_1, v_0), \quad \text{and} \quad W := w,$$

we conclude that (2.10)–(2.12) take the form of (2.4)–(2.6). Therefore, it is natural to interpret the problem (2.10) as the dual (2.11).

The assumptions (H1)–(H3) are satisfied. Indeed, (H1) was already verified in Section 2.1.2, while (H2) follows again from a usual trace theorem. Finally, (H3) is equivalent to the following *hidden regularity* theorem of Lions [96], [97]:[8]

Theorem 2.7. *The solutions of* (2.11) *satisfy the* direct *inequality*

$$\int_I \int_\Gamma \left| \frac{\partial u}{\partial \nu} \right|^2 d\Gamma \, dt \leq c \int_\Omega |\nabla u_0|^2 + |\nabla \Delta^{-1} u_1|^2 \, dx$$

for every time interval I, with a constant c depending only on the length $|I|$.

Applying Theorem 2.4, we obtain the following proposition:

Proposition 2.8. *Given*

$$v_0 \in H_0^1(\Omega), \quad v_1 \in H^{-1}(\Omega), \quad and \quad w \in L^2(\mathbb{R}; (L^2(\Gamma)'))$$

arbitrarily, the problem (2.10) *has a unique solution satisfying*

$$v \in C(\mathbb{R}; H_0^1(\Omega)) \cap C^1(\mathbb{R}; H^{-1}(\Omega)).$$

2.3 Observability and Controllability

In this section we review an important duality principle between observability and controllability. Again, we begin by presenting an abstract framework, followed by examples.

Let us return to the abstract dual problems

$$U' = \mathcal{A}U, \quad U(0) = U_0, \tag{2.13}$$

and

$$V' = -\mathcal{A}^* V + \mathcal{B}^* W, \quad V(0) = V_0, \tag{2.14}$$

of the preceding section, satisfying hypotheses (H1)–(H3) (pp. 10, 13). We are going to show that there is an intimate relation between the observability of the first and the controllability of the second equation.

Let us also assume the following *inverse* inequality to (H3):

(H4) There exists a bounded interval I' and a positive number c' such that the solutions of (2.13) satisfy the inequality

$$\|U_0\|_{\mathcal{H}} \leq c' \|\mathcal{B}U\|_{L^2(I';\mathcal{G})}$$

for all $U_0 \in \mathcal{Z}$.

---

[8]See also [67], p. 29.

Remarks.

- This is also called an *observability* inequality, because it implies that two different initial data U_0 in the problem (2.13) lead to different observations $\mathcal{B}U|_{I'}$, so that (applying the inverse inequality to the difference of the two solutions, which is also a solution of our *linear* equation) the observation is sufficient in order to distinguish the unknown initial data.
- Unlike condition (H3), in (H4) the length of the interval I is important: it is related to the critical time of observability. Indeed, an elementary argument based on the time invariance of equation (2.13) and on the estimate (2.5) of Theorem 2.1 (p. 10) shows that if (H4) is satisfied for some interval I', then it is also satisfied for every translate of this interval, with perhaps a different constant c'. It follows that the inequality also holds with some constant c' for every interval having *at least* the same length. Thus there exists a number $T_0 \geq 0$ such that the observability inequality holds for *all* intervals longer than T_0 and for *no* intervals shorter than T_0. (In this abstract setting we cannot draw any conclusion regarding intervals of length *equal* to T_0.) Note that we do not exclude the case $T_0 = 0$: then the observability inequality holds for all intervals. (But the constant c' explodes as the length of I' tends to zero.)

Now the main result of this section is the following:

Theorem 2.9. *Assume (H1)–(H4) and let $T > |I'|$. Then for every initial state $V_0 \in \mathcal{H}'$ there exists a function[9] $W \in L^2(0,T;\mathcal{G}')$ such that the solution of (2.14) satisfies the final condition $V(T) = 0$. (We say that the control W drives the system to rest in time T.)*

Moreover, we can choose W satisfying

$$\|W\|_{L^2(0,T;\mathcal{G}')} \leq c_T\|V_0\|_{\mathcal{H}'} \qquad (2.15)$$

with a constant c_T independent of the particular choice of $V_0 \in \mathcal{H}'$.

Proof. As a consequence of hypotheses (H1)–(H4) the formula

$$(U_0, \tilde{U}_0) \mapsto \int_0^T (\mathcal{B}e^{t\mathcal{A}}U_0, \mathcal{B}e^{t\mathcal{A}}\tilde{U}_0)_{\mathcal{G}}\, dt$$

defines a continuous, symmetric, and coercive bilinear form in $\mathcal{H}$. Applying the Riesz–Fréchet theorem, we see that there exists a self-adjoint, positive definite isomorphism $\Lambda \in L(\mathcal{H}, \mathcal{H}')$ such that

$$\langle \Lambda U_0, \tilde{U}_0 \rangle_{\mathcal{H}',\mathcal{H}} = \int_0^T (\mathcal{B}e^{t\mathcal{A}}U_0, \mathcal{B}e^{t\mathcal{A}}\tilde{U}_0)_{\mathcal{G}}\, dt$$

---

[9]We extend the function W for example by 0 outside $[0, T]$ so as to obtain a function $W \in L^2_{\text{loc}}(\mathbb{R}; \mathcal{G}')$. It is called a *control* function.

for all $U_0, \tilde{U}_0 \in \mathcal{H}$.

Let us denote by $J : \mathcal{G} \to \mathcal{G}'$ the canonical Riesz isomorphism. Given $V_0 \in \mathcal{H}'$ arbitrarily, we claim that the control

$$W(s) := -J\mathcal{B}e^{t\mathcal{A}}\Lambda^{-1}V_0$$

drives V_0 to rest in time T. Indeed, for any given $U_0 \in \mathcal{H}$, using (2.6) (p. 14) we have

$$
\begin{aligned}
\langle V(T), U(T) \rangle_{\mathcal{H}',\mathcal{H}} &= \langle V_0, U_0 \rangle_{\mathcal{H}',\mathcal{H}} + \int_0^T \langle W(t), \mathcal{B}U(t) \rangle_{\mathcal{G}',\mathcal{G}} \, dt \\
&= \langle V_0, U_0 \rangle_{\mathcal{H}',\mathcal{H}} - \int_0^T (\mathcal{B}e^{t\mathcal{A}}\Lambda^{-1}V_0, \mathcal{B}e^{t\mathcal{A}}U_0)_{\mathcal{G}} \, dt \\
&= \langle V_0, U_0 \rangle_{\mathcal{H}',\mathcal{H}} - \langle \Lambda\Lambda^{-1}V_0, U_0 \rangle_{\mathcal{H}',\mathcal{H}} \\
&= 0.
\end{aligned}
$$

Since $e^{T\mathcal{A}}$ is an automorphism, $U(T)$ runs over the whole of $\mathcal{H}$ if U_0 does. Therefore, we conclude that $V(T) = 0$.

Finally, the estimate of the norm of the control results from a direct computation, using the isomorphic character of J, Λ, and the *direct inequality*

$$
\begin{aligned}
\|W\|_{L^2(0,T;\mathcal{G}')} &= \|J\mathcal{B}e^{s\mathcal{A}}\Lambda^{-1}V_0\|_{L^2(0,T;\mathcal{G}')} \\
&= \|\mathcal{B}e^{s\mathcal{A}}\Lambda^{-1}V_0\|_{L^2(0,T;\mathcal{G})} \\
&\leq c_1 \|\Lambda^{-1}V_0\|_{\mathcal{H}} \\
&\leq c_2 \|V_0\|_{\mathcal{H}'}.
\end{aligned}
$$

Remark. As a matter of fact, under hypotheses (H1)–(H3) the controllability of (2.14) is *equivalent* to the observability of (2.13). Indeed, fix $U_0 \in \mathcal{Z}$ arbitrarily. For every $V_0 \in \mathcal{H}'$ choose an exact control W satisfying the norm inequality (2.15). Using the definition (2.6) of weak solutions, the corresponding solutions of (2.13) and (2.14) satisfy the equality

$$\langle V_0, U_0 \rangle_{\mathcal{H}',\mathcal{H}} = -\int_0^T \langle W(t), \mathcal{B}U(t) \rangle_{\mathcal{G}',\mathcal{G}} \, dt.$$

Using (2.15), it follows that

$$|\langle V_0, U_0 \rangle_{\mathcal{H}',\mathcal{H}}| \leq c_T \|V_0\|_{\mathcal{H}'} \|\mathcal{B}U\|_{L^2(0,T;\mathcal{G})}.$$

Since this is true for all $V_0 \in \mathcal{H}'$, using the Hahn–Banach theorem we conclude that

$$\|U_0\|_{\mathcal{H}} \leq c_T \|\mathcal{B}U\|_{L^2(I';\mathcal{G})}.$$

We note that this duality relation remains valid if we assume instead of (H1) only that $\mathcal{A}$ generates a *semigroup* in $\mathcal{H}$; see Dolecki and Russell [32].

In [96], [97] Lions developed a general and systematic approach for the study of exact controllability of linear distributed systems, the so-called Hilbert uniqueness method (HUM). It was based on the duality principle discussed in this section. Since these references contain a great number of examples (see also [67] for a textbook exposition), in the present book we restrict ourselves to the study of observability, leaving to the reader the formulation of the corresponding controllability results. Let us just recall two applications.

2.3.1 Wave Equation with Dirichlet Control

We return to the problems studied in Section 2.2.1 (p. 15):

$$\begin{cases} v'' - \Delta v = 0 & \text{in } \mathbb{R} \times \Omega, \\ v = w & \text{on } \mathbb{R} \times \Gamma, \\ v(0) = v_0 \quad \text{and} \quad v'(0) = v_1 & \text{in } \Omega; \end{cases} \tag{2.16}$$

and

$$\begin{cases} u'' - \Delta u = 0 & \text{in } \mathbb{R} \times \Omega, \\ u = 0 & \text{on } \mathbb{R} \times \Gamma, \\ u(0) = u_0 \quad \text{and} \quad u'(0) = u_1 & \text{in } \Omega, \end{cases} \tag{2.17}$$

with the observation of $\partial_\nu u$.

We recall that hypotheses (H1)–(H3) are satisfied. Solving a conjecture of Lions [94], L.F. Ho [50] proved that hypothesis (H4) is also satisfied. This was subsequently improved[10] by Lions [96], [97], who weakened the assumption on the length of I:

Theorem 2.10. *Let R denote the radius of the smallest open ball containing Ω. The solutions of (2.17) satisfy the* inverse *inequality*

$$\int_\Omega |\nabla u_0|^2 + |u_1|^2 \, dx \le c \int_I \int_\Gamma \left| \frac{\partial u}{\partial \nu} \right|^2 \, d\Gamma \, dt$$

for all intervals I of length $|I| > 2R$, with a constant c depending only on $|I|$.

Remarks.

- The above-mentioned proofs were based on the multiplier method. The critical length for the validity of the inverse inequality was determined by Bardos, Lebeau, and Rauch [11], using microlocal analysis.
- The inverse inequality cannot hold for arbitrarily short intervals because of the finite propagation property of the wave equation; see, e.g., Remark 3.6 in [67] (p. 40) for a short proof.

---

[10]See also [62] or [67] (p. 36) for a simplification of his proof.

- If we consider only initial data with $u_0 = 0$, then the above inverse inequality also holds for the shorter intervals $I = (0, T)$ with $T > R$. This easily follows from the theorem by observing that the solutions are *odd* functions of the time t, so that

$$\int_0^T \int_\Gamma \left|\frac{\partial u}{\partial \nu}\right|^2 d\Gamma \, dt = \frac{1}{2} \int_{-T}^T \int_\Gamma \left|\frac{\partial u}{\partial \nu}\right|^2 d\Gamma \, dt,$$

  and the length of the interval $(-T, T)$ on the right-hand side is greater than $2R$.
- An analogous result holds if we consider only initial data with $u_1 = 0$: now the solutions are *even* functions of the time t.

Applying Theorem 2.9, we obtain the following result of Lions [96], [97]:

Theorem 2.11. *If $T > 2R$, then for any given initial data $u_0 \in L^2(\Omega)$ and $u_1 \in H^{-1}(\Omega)$ there exists a function $w \in L^2(0, T; L^2(\Omega))$ such that the solution of* (2.16) *satisfies*

$$v(T) = v'(T) = 0 \quad in \quad \Omega.$$

Remark. Applying the last two remarks above, we obtain that if $T > R$, then for every initial data $u_0 \in L^2(\Omega)$ and $u_1 \in H^{-1}(\Omega)$ there exists a function $w \in L^2(0, T; L^2(\Omega))$ such that the solution of (2.16) satisfies

$$v(T) = 0 \quad in \quad \Omega,$$

and another function $w \in L^2(0, T; L^2(\Omega))$ such that the solution of (2.16) satisfies

$$v'(T) = 0 \quad in \quad \Omega.$$

In other words, half the time is sufficient if we want to control only the state or the velocity, but not both. This was proved earlier by Lions using a different argument in [97], pp. 95–102.

2.3.2 A Petrovsky System

We return to the problems

$$\begin{cases} v'' + \Delta^2 v = 0 & \text{in} \quad \mathbb{R} \times \Omega, \\ v = 0 \quad \text{and} \quad \Delta v = w & \text{on} \quad \mathbb{R} \times \Gamma, \\ v(0) = v_0 \quad \text{and} \quad v'(0) = v_1 & \text{in} \quad \Omega, \end{cases} \tag{2.18}$$

and

$$\begin{cases} u'' + \Delta^2 u = 0 & \text{in} \quad \mathbb{R} \times \Omega, \\ u = \Delta u = 0 & \text{on} \quad \mathbb{R} \times \Gamma, \\ u(0) = u_0 \quad \text{and} \quad u'(0) = u_1 & \text{in} \quad \Omega, \end{cases} \tag{2.19}$$

of Section 2.2.2 (p. 16).

We recall that hypotheses (H1)–(H3) are satisfied. In [96] and [97] Lions proved that hypothesis (H4) is also satisfied if the interval I is sufficiently long.[11] Subsequently, Zuazua [139] proved that the inverse inequality holds in fact for all intervals:[12]

Theorem 2.12. *The solutions of* (2.19) *satisfy the* inverse *inequality*

$$\int_\Omega |\nabla u_0|^2 + |\nabla \Delta^{-1} u_1|^2 \, dx \le c \int_I \int_\Gamma \left| \frac{\partial u}{\partial \nu} \right|^2 \, d\Gamma \, dt$$

for every time interval I, with some constant c depending only on the length $|I|$ of I.

Remark. All these proofs used the multiplier method. The possibility of taking arbitrarily short intervals is related to the infinite propagation speed in this equation.

Applying Theorem 2.9, we deduce the following improvement[13] of earlier results of Lions [96], [97] and Zuazua [139]:

Theorem 2.13. *For arbitrary $T > 0$, $u_0 \in H_0^1(\Omega)$, and $u_1 \in H^{-1}(\Omega)$, there exists a function $w \in L^2(0, T; L^2(\Omega))$ such that the solution of* (2.18) *satisfies*

$$v(T) = v'(T) = 0 \quad in \quad \Omega.$$

2.4 Observability and Stabilizability

We present here an approach to the uniform stabilization, introduced in [70], analogous to HUM. This leads to the construction of boundary feedbacks with *arbitrarily large* decay rates.

Let us return to our abstract framework,

$$U' = \mathcal{A}U, \quad U(0) = U_0, \quad \psi = \mathcal{B}U,$$
$$V' = -\mathcal{A}^*V + \mathcal{B}^*W, \quad V(0) = V_0, \tag{2.20}$$

and assume hypotheses (H1)–(H4) again (pp. 10, 13, 18). Fix two numbers $T > |I'|$, $\omega > 0$, set $T_\omega = T + (2\omega)^{-1}$, define[14]

$$e_\omega(s) = \begin{cases} e^{-2\omega s} & \text{if } 0 \le s \le T, \\ 2\omega e^{-2\omega T}(T_\omega - s) & \text{if } T \le s \le T_\omega, \end{cases}$$

---

[11] His proof was simplified in [62].

[12] See also [67] for a constructive proof of this result.

[13] See [67], p. 83. In the former results two controls were used.

[14] This particular weight function was proposed by Bourquin [16].

and set

$$\langle \Lambda_\omega U_0, \tilde{U}_0 \rangle_{\mathcal{H}', \mathcal{H}} := \int_0^{T_\omega} e_\omega(s)(\mathcal{B}e^{s\mathcal{A}}U_0, \mathcal{B}e^{s\mathcal{A}}\tilde{U}_0)_{\mathcal{G}} \, ds.$$

Then Λ_ω is a self-adjoint, positive definite isomorphism $\Lambda_\omega \in L(\mathcal{H}, \mathcal{H}')$. Let us denote by $J : \mathcal{G} \to \mathcal{G}'$ the canonical Riesz anti-isomorphism.

The following result is a special case of a theorem obtained in [70].

Theorem 2.14. *Assume (H1)–(H4) and fix $\omega > 0$ arbitrarily. Then the problem*

$$V' = (-\mathcal{A}^* - \mathcal{B}^* J \mathcal{B} \Lambda_\omega^{-1})V, \qquad V(0) = V_0, \tag{2.21}$$

is well-posed in $\mathcal{H}'$. Furthermore, there exists a constant M such that the solutions of (2.21) satisfy the estimates

$$\|V(t)\|_{\mathcal{H}'} \leq M\|V_0\|_{\mathcal{H}'}e^{-\omega t} \tag{2.22}$$

for all $V_0 \in \mathcal{H}'$ and for all $t \geq 0$.

In other words, this theorem asserts that the *feedback law*

$$W = -J\mathcal{B}\Lambda_\omega^{-1}V$$

uniformly stabilizes the control problem (2.20) with a decay rate at least equal to ω.

The well-posedness means here that (2.21) has a unique solution $V \in C(\mathbb{R}; \mathcal{H}')$ for every $V_0 \in \mathcal{H}'$.

Sketch of the proof. We admit the well-posedness of (2.21), and we write Λ_ω in the following form:

$$\Lambda_\omega = \int_0^{T_\omega} e_\omega(s)e^{s\mathcal{A}^*}\mathcal{B}^* J\mathcal{B}e^{s\mathcal{A}} \, ds.$$

Fix $V_0 \in \mathcal{H}'$ arbitrarily and consider the solution of (2.21). A simple (formal) computation leads to the following identity:

$$\frac{d}{dt}\langle \Lambda_\omega^{-1}V, V \rangle_{\mathcal{H}, \mathcal{H}'} = \langle \Lambda_\omega^{-1}V, (-\mathcal{A}^*\Lambda_\omega - \Lambda_\omega\mathcal{A} - 2\mathcal{B}^* J\mathcal{B})\Lambda_\omega^{-1}V \rangle_{\mathcal{H}, \mathcal{H}'}. \tag{2.23}$$

Since

$$2\omega e_\omega(s) \leq -e_\omega'(s) \quad \text{and} \quad e_\omega(T_\omega) = 0,$$

we have

$$-\mathcal{A}^*\Lambda_\omega - \Lambda_\omega\mathcal{A} + 2\omega\Lambda_\omega \leq -\int_0^{T_\omega} \frac{d}{ds}\left(e_\omega(s)e^{s\mathcal{A}^*}\mathcal{B}^* J\mathcal{B}e^{s\mathcal{A}}\right) \, ds = \mathcal{B}^* J\mathcal{B}.$$

Hence we obtain that

$$-\mathcal{A}^*\Lambda_\omega - \Lambda_\omega\mathcal{A} - 2\mathcal{B}^* J\mathcal{B} \leq -2\omega\Lambda_\omega.$$

(This means that the right-hand side minus the left-hand side is positive semidefinite.) Therefore, we deduce from the identity (2.23) the following inequality:

$$\frac{d}{dt} \langle \Lambda_\omega^{-1} V, V \rangle_{\mathcal{H}, \mathcal{H}'} \leq -2\omega \langle \Lambda_\omega^{-1} V, V \rangle_{\mathcal{H}, \mathcal{H}'}.$$

Hence

$$\langle \Lambda_\omega^{-1} V(t), V(t) \rangle_{\mathcal{H}, \mathcal{H}'} \leq \langle \Lambda_\omega^{-1} V_0, V_0 \rangle_{\mathcal{H}, \mathcal{H}'} e^{-2\omega t} \qquad (2.24)$$

for all $t \geq 0$. Since $\Lambda_\omega \in L(\mathcal{H}, \mathcal{H}')$ is a self-adjoint, positive definite isomorphism, there exist two constants $c_1, c_2 > 0$ such that

$$c_1 \|V\|_H^2 \leq \langle \Lambda_\omega^{-1} V, V \rangle_{\mathcal{H}, \mathcal{H}'} \leq c_2 \|V\|_H^2$$

for all $V \in \mathcal{H}'$. Using these inequalities, (2.24) implies (2.22) with $M = \sqrt{c_2/c_1}$.

The above proof is correct in the finite-dimensional case, but there are some technical difficulties in the infinite-dimensional case due to the rather weak regularity of the solutions of (2.21). See [70] for the complete proof in the general case.

Remark. Various numerical and experimental tests were conducted by Bourquin and his collaborators Briffaut, Collet, Ratier, and Urquiza on the efficiency of these feedbacks: see, e.g., [17], [18], [21], [116], [134].

We give only two applications of this theorem, and we refer to [70] and [71] for further ones.

2.4.1 Wave Equation with Dirichlet Feedback

We recall from Section 2.2.1 (p. 15) that if we write the problem

$$\begin{cases} u'' - \Delta u = 0 & \text{in } \mathbb{R} \times \Omega, \\ u = 0 & \text{on } \mathbb{R} \times \Gamma, \\ u(0) = u_0 \quad \text{and} \quad u'(0) = u_1 & \text{in } \Omega, \\ \psi = \partial_\nu u & \text{on } \mathbb{R} \times \Gamma, \end{cases}$$

in the abstract form

$$U' = \mathcal{A}U, \quad U(0) = U_0, \qquad \psi = \mathcal{B}U,$$

then the corresponding control problem

$$V' = -\mathcal{A}^* V + \mathcal{B}^* W, \quad V(0) = V_0,$$

is equivalent to

$$\begin{cases} v'' - \Delta v = 0 & \text{in } \mathbb{R} \times \Omega, \\ v = w & \text{on } \mathbb{R} \times \Gamma, \\ v(0) = v_0 \quad \text{and} \quad v'(0) = v_1 & \text{in } \Omega. \end{cases}$$

Since hypotheses (H1)–(H4) are satisfied, we can apply Theorem 2.14. It remains to identify the feedback $W = -J\mathcal{B}\Lambda_\omega^{-1}V$. Writing the operator

$$\Lambda_\omega^{-1} : H^{-1}(\Omega) \times L^2(\Omega)' \to H_0^1(\Omega) \times L^2(\Omega)$$

in the matrix form

$$\Lambda_\omega^{-1} = \begin{pmatrix} P & -Q \\ -R & S \end{pmatrix}$$

and using the definition of $\mathcal{B}$, we have

$$w = -J\mathcal{B}\Lambda_\omega^{-1}x = \frac{\partial}{\partial_\nu}(Py' + Qy)$$

if we identify $\mathcal{G} = L^2(\Gamma)$ with its dual $\mathcal{G}'$ as usual. We have thus proved the following result:

Theorem 2.15. *Fix an arbitrarily large positive number ω. Then there exist two bounded linear maps*

$$P : H^{-1}(\Omega) \to H_0^1(\Omega), \qquad Q : L^2(\Omega) \to H_0^1(\Omega),$$

and a constant M such that the closed-loop problem

$$\begin{cases} v'' - \Delta v = 0 & \text{in} \quad \mathbb{R} \times \Omega, \\ v = \partial_\nu(Pv' + Qv) & \text{on} \quad \mathbb{R} \times \Gamma, \\ v(0) = v_0 \quad \text{and} \quad v'(0) = v_1 & \text{in} \quad \Omega, \end{cases}$$

is well-posed in $\mathcal{H} := L^2(\Omega)' \times H^{-1}(\Omega)$, and its solutions satisfy the estimates

$$\|(v, v')(t)\|_{\mathcal{H}} \leq M\|(v_0, v_1)\|_{\mathcal{H}} e^{-\omega t}$$

for all $t \geq 0$ and for all $(v_0, v_1) \in \mathcal{H}$.

2.4.2 A Petrovsky System with Hinged Boundary Conditions

Applying the results of Sections 2.1.2, 2.2.2 and 2.3.2 (pp. 12, 16, and 22), we can show that Theorem 2.14 yields the following theorem:

Theorem 2.16. *Fix an arbitrarily large positive number ω. Then there exist two bounded linear maps*

$$P : H^{-1}(\Omega) \to H_0^1(\Omega), \qquad Q : H_0^1(\Omega) \to H_0^1(\Omega),$$

and a constant M such that the closed-loop problem

$$\begin{cases} v'' + \Delta^2 v = 0 & \text{in} \quad \mathbb{R} \times \Omega, \\ v = 0 \quad \text{and} \quad \Delta v = \partial_\nu(Pv' + Qv) & \text{on} \quad \mathbb{R} \times \Gamma, \\ v(0) = v_0 \quad \text{and} \quad v'(0) = v_1 & \text{in} \quad \Omega, \end{cases}$$

is well-posed in $\mathcal{H} := H_0^1(\Omega) \times H^{-1}(\Omega)$, *and its solutions satisfy the estimates*

$$\|(v, v')(t)\|_{\mathcal{H}} \leq M\|(v_0, v_1)\|_{\mathcal{H}} e^{-\omega t}$$

for all $t \geq 0$ *and for all* $(v_0, v_1) \in \mathcal{H}$.

The proof is left to the reader (or see [70]).

2.5 Partial Observation, Control, and Stabilization

In applications it is often desirable to generalize Theorems 2.9 and 2.14 (pp. 19, 24) for several reasons:

- Sometimes the system

$$U' = \mathcal{A}U, \quad U(0) = U_0, \tag{2.25}$$

  is only *partially* observable; i.e., only a weakened form of hypothesis (H4) is satisfied.
- Sometimes the system

$$V' = -\mathcal{A}^*V + \mathcal{B}^*W, \quad V(0) = V_0, \tag{2.26}$$

  is only *partially* controllable; i.e., not all initial states can be steered to zero.
- Sometimes the system (2.26) is not stabilizable, but some initial states can still be driven to zero by an appropriate feedback.
- Even when stabilizable, it may be too expensive to stabilize the whole state: it could be more economic and at the same time completely acceptable from the point of view of applications to stabilize only a finite number of modes.[15]

Let us consider again the abstract framework of the problems (2.25) and (2.26) as in Sections 2.2 and 2.3 (pp. 13, 18). We continue to assume that hypotheses (H1)–(H3) (pp. 10, 13) are satisfied. Then for any fixed continuous (strictly) positive function f given on some interval $[0, T]$, the formula

$$\langle \Lambda_f U_0, \tilde{U}_0 \rangle := \int_0^T (\mathcal{B}e^{t\mathcal{A}}U_0, \mathcal{B}e^{t\mathcal{A}}\tilde{U}_0)_\mathcal{G} \, dt \tag{2.27}$$

defines a continuous linear map

$$\Lambda_f : \mathcal{H} \to \mathcal{H}'.$$

Let us take a closer look at Λ_f:

---

[15]This is exactly what has been done in the numerical and physical experiments carried over by Bourquin et al. [17], [18], [21], [116], [134].

Lemma 2.17. *The map Λ_f has the following properties:*

(a) $\Lambda_f^* = \Lambda_f$;

(b) $\langle \Lambda_f U_0, U_0 \rangle \geq 0$ *for all* $U_0 \in \mathcal{H}$;

(c) $\langle \Lambda_f U_0, U_0 \rangle = 0$ *if and only if* $\Lambda_f U_0 = 0$;

(d) $N(\Lambda_f) = R(\Lambda_f)^{\perp}$.

In the last assertion, $N(\Lambda_f)$ and $R(\Lambda_f)$ denote the kernel (or nullset) and the range of Λ_f, respectively, and $R(\Lambda_f)^{\perp}$ denotes the orthogonal complement of $R(\Lambda_f) \subset \mathcal{H}'$ in $\mathcal{H}$.

Proof. Assertions (a), (b), and the inverse implication in (c) follow at once from the definition of Λ_f. The direct implication in (c) is a consequence of the generalized Cauchy–Schwarz inequality

$$|\langle \Lambda_f U_0, \tilde{U}_0 \rangle|^2 \leq \langle \Lambda_f U_0, U_0 \rangle \cdot \langle \Lambda_f \tilde{U}_0, \tilde{U}_0 \rangle,$$

which holds for every positive *semidefinite* quadratic form.

Turning to the proof of (d), if $U_0 \in \mathcal{H}$ and $\tilde{U}_0 \in N(\Lambda_f)$, then

$$\langle \Lambda_f U_0, \tilde{U}_0 \rangle = \langle U_0, \Lambda_f \tilde{U}_0 \rangle = \langle U_0, 0 \rangle = 0.$$

This proves the inclusion

$$N(\Lambda_f) \subset R(\Lambda_f)^{\perp}.$$

On the other hand, if $\tilde{U}_0 \in R(\Lambda_f)^{\perp}$, then we have

$$\langle \Lambda_f \tilde{U}_0, U_0 \rangle = \langle \tilde{U}_0, \Lambda_f U_0 \rangle = 0$$

for every $U_0 \in \mathcal{H}$. This proves the inverse inclusion

$$R(\Lambda_f)^{\perp} \subset N(\Lambda_f).$$

Remarks.

- It follows from property (c) that $N(\Lambda_f)$ is the set of *nonobservable* initial states for problem (2.25) in the sense that the observation of $\mathcal{B}e^{tA}U_0$ on $[0, T]$ does not allow us to distinguish U_0 from 0.
- In particular, this also shows that $N(\Lambda_f)$ does not depend on the particular choice of the function f. Since

$$\overline{R(\Lambda_f)} = R(\Lambda_f)^{\perp\perp} = N(\Lambda_f)^{\perp},$$

the *closure* of the range $R(\Lambda_f)$ of Λ_f does not depend on f either.

We are going to show that $R(\Lambda_f)$ also has a natural control-theoretical interpretation. Let us adopt the following definition:

Definition. A state $V_0 \in \mathcal{H}'$ is *controllable* (in time T) if there exists a "control function" $W \in L^2(0, T; \mathcal{G}')$ such that the solution of (2.26) satisfies $V(T) = 0$.

We need the following weakening of hypothesis (H4) (p. 18):

(H4') There exists a constant $c' > 0$ such that the solutions of (2.25) satisfy the inequality

$$\inf_{\tilde{U}_0 \in N(\Lambda_f)} \|U_0 + \tilde{U}_0\|_{\mathcal{H}}^2 \le c' \int_0^T f(t) \|\mathcal{B}U(t)\|_{\mathcal{G}}^2 \, dt$$

for all $U_0 \in \mathcal{Z}$.

If Λ_f is one-to-one, then this hypothesis reduces to (H4), and then Λ_f is in fact an isomorphism of $\mathcal{H}$ onto $\mathcal{H}'$. In the general case, we have the following lemma:

Lemma 2.18. *Assume (H1)–(H3) and (H4'). Then Λ_f has a closed range $R(\Lambda_f)$ in $\mathcal{H}'$. Moreover, the quotient map of Λ_f with respect to its kernel $N(\Lambda_f)$ is an isomorphism of $\mathcal{H}/N(\Lambda_f)$ onto*

$$R(\Lambda_f) = N(\Lambda_f)^\perp.$$

Proof. It follows from hypotheses (H1)–(H3) and (H4') that the formula (2.27) defines an equivalent scalar product on the factor space $\mathcal{H}/N(\Lambda_f)$. Applying the Riesz–Fréchet theorem, we conclude that the quotient map of Λ_f with respect to its kernel $N(\Lambda_f)$ is an isomorphism of $\mathcal{H}/N(\Lambda_f)$ onto the dual space of $\mathcal{H}/N(\Lambda_f)$, i.e., onto $N(\Lambda_f)^\perp$; see, e.g., Rudin [121], Theorem 4.9.

Now we prove the following important theorem:

Theorem 2.19. *Assume (H1)–(H3) and (H4'). Then $R(\Lambda_f) = N(\Lambda_f)^\perp$ is the set of controllable states for the problem (2.26).*

Proof. If V_0 is a controllable state, then it is orthogonal to $N(\Lambda_f)$, so that V_0 belongs to $R(\Lambda_f)$ by the equality just proved. Indeed, if $W : [0, T] \to \mathcal{G}'$ is a suitable control for V_0, then we have for every $U_0 \in N(\Lambda_f)$ the equality

$$0 = \int_0^T \langle U, V' + \mathcal{A}^*V + \mathcal{B}^*W \rangle \, dt$$

$$= \langle U(T), V(T) \rangle - \langle U_0, V_0 \rangle + \int_0^T \langle -U' + \mathcal{A}U, V \rangle \, dt + \int_0^T \langle \mathcal{B}U, W \rangle \, dt$$

$$= -\langle U_0, V_0 \rangle,$$

because the first two integrals vanish by (2.25) and (2.26), $V(T) = 0$ by the choice of W, and $\mathcal{B}U = 0$ a.e. on $[0, T]$ because U_0 is a nonobservable state. Thus V_0 is orthogonal to every $U_0 \in N(\Lambda_f)$.

Conversely, every $V_0 \in R(\Lambda_f)$ is controllable. Indeed, choose $U_0 \in \mathcal{H}$ such that $\Lambda_f U_0 = V_0$ (this is possible because Λ_f is surjective by the preceding lemma) and consider the control

$$W(t) := f(t) J \mathcal{B} U(t),$$

where U is the solution of (2.25) and $J : \mathcal{G} \to \mathcal{G}'$ is the canonical Riesz isomorphism. Now, if $\tilde{U}_T \in \mathcal{H}$ and $\tilde{U}$ denotes the solution of the problem

$$\tilde{U}' = \mathcal{A}\tilde{U}, \quad \tilde{U}(T) = U_T, \tag{2.28}$$

then we have the following equality:

$$
\begin{aligned}
0 &= \int_0^T \langle \tilde{U}, V' + \mathcal{A}^* V + \mathcal{B}^* W \rangle \, dt \\
&= \langle \tilde{U}_T, V(T) \rangle - \langle \tilde{U}(0), V_0 \rangle + \int_0^T \langle -\tilde{U}' + \mathcal{A}\tilde{U}, V \rangle \, dt + \int_0^T \langle \mathcal{B}\tilde{U}, W \rangle \, dt \\
&= \langle \tilde{U}_T, V(T) \rangle.
\end{aligned}
$$

Indeed, the first two integrals vanish by (2.26) and (2.28), and we have

$$\int_0^T \langle \mathcal{B}\tilde{U}, W \rangle \, dt = \int_0^T f(t)(\mathcal{B}\tilde{U}(t), \mathcal{B}U(t))_\mathcal{G} \, dt = \langle \tilde{U}(0), \Lambda_f U_0 \rangle = \langle \tilde{U}(0), V_0 \rangle.$$

Thus $V(T)$ is orthogonal to every $\tilde{U}_T \in \mathcal{H}$, so that $V(T) = 0$.

Next we have the following generalization of Theorem 2.14 (p. 24). Let us introduce the same operator Λ_ω as there.[16]

Theorem 2.20. *Assume (H1)–(H3) and (H4′), and fix $\omega > 0$ arbitrarily. Then the problem*

$$V' = (-\mathcal{A}^* - \mathcal{B}^* J \mathcal{B} \Lambda_\omega^{-1}) V, \qquad V(0) = V_0, \tag{2.29}$$

is well-posed in $R(\Lambda_\omega) = N(\Lambda_\omega)^\perp$. Furthermore, there exists a constant M such that the solutions of (2.29) satisfy the estimates

$$\|V(t)\|_{\mathcal{H}'} \leq M \|V_0\|_{\mathcal{H}'} e^{-\omega t}$$

for all $V_0 \in R(\Lambda_\omega)$ and for all $t \geq 0$.

Proof. For simplicity we prove the theorem only in the finite-dimensional case.

First we show that $N(\Lambda_\omega)$ is an invariant subspace of $\mathcal{A}$. Indeed, if U_0 belongs to $N(\Lambda_\omega)$, then $\mathcal{B}\mathcal{A}^k U_0 = 0$ for all $k = 0, 1, \ldots$. Hence $\mathcal{B}\mathcal{A}^k \mathcal{A} U_0 = 0$ for all $k = 0, 1, \ldots$, whence $\mathcal{A}U_0 \in N(\Lambda_\omega)$.

---

[16]We should write Λ_{e_ω} instead of Λ_ω, but we prefer to keep the earlier notation.

Next we prove that $R(\Lambda_\omega)$ is an invariant subspace of the operator

$$-\mathcal{A}^* - \mathcal{B}^* J \mathcal{B} \Lambda_\omega^{-1}.$$

Indeed, first of all,

$$(\mathcal{A}^* + \mathcal{B}^* J \mathcal{B} \Lambda_\omega^{-1}) V_0 = (\mathcal{A}^* \Lambda_\omega + \mathcal{B}^* J \mathcal{B}) U_0$$

is well-defined for all $V_0 = \Lambda_\omega U_0 \in R(\Lambda_\omega)$; i.e., its value does not depend on the particular choice of U_0. This follows from the fact that if $\tilde{U}_0 \in N(\Lambda_\omega)$, then $\mathcal{B}\tilde{U}_0 = 0$.

It remains to show that

$$(\mathcal{A}^* \Lambda_\omega + \mathcal{B}^* J \mathcal{B}) U_0 \perp N(\Lambda_\omega).$$

If $\tilde{U}_0 \in N(\Lambda_\omega)$, then

$$
\begin{aligned}
\langle (\mathcal{A}^* \Lambda_\omega + \mathcal{B}^* J \mathcal{B}) U_0, \tilde{U}_0 \rangle &= \langle \Lambda_\omega U_0, \mathcal{A}\tilde{U}_0 \rangle + \langle \mathcal{B}U_0, \mathcal{B}\tilde{U}_0 \rangle \\
&= \langle U_0, \Lambda_\omega \mathcal{A}\tilde{U}_0 \rangle + \langle \mathcal{B}U_0, \mathcal{B}\tilde{U}_0 \rangle \\
&= 0
\end{aligned}
$$

because $\Lambda_\omega \mathcal{A}\tilde{U}_0 = 0$ by the $\mathcal{A}$-invariance of $N(\Lambda_\omega)$ and $\mathcal{B}\tilde{U}_0 = 0$ by the condition $\tilde{U}_0 \in N(\Lambda_\omega)$.

It follows from what we proved that the problem (2.29) is well-posed in $R(\Lambda_\omega)$. We may now repeat the proof of Theorem 2.14.

Well-Posedness in a Riesz Basis Setting

In many problems of practical interest, the infinitesimal generator of the semi-group has an additional property: it is diagonalizable in some simple sense. This aspect simplifies the structure of the solutions, and also turns to be useful when we study the observability properties of these systems. In this chapter we prove a general existence theorem for such operators, and we give a large number of examples where this theorem applies.

3.1 An Abstract Existence Theorem

We are going to investigate the well-posedness of the problem

$$U' = \mathcal{A}U, \quad U(0) = U_0, \tag{3.1}$$

in a *complex* Hilbert space $\mathcal{H}$, where $\mathcal{A}$ is a (bounded or unbounded) linear operator defined on some linear subspace of $\mathcal{H}$, with values in $\mathcal{H}$.

If $\mathcal{H}$ is finite-dimensional, then a classical theorem of Jordan ensures the existence of a basis formed by ordinary and generalized eigenvectors of $\mathcal{A}$. More precisely, there exist a basis

$$\{E_{k,\ell} \ : \ k = 1, \dots, K, \quad \ell = 1, \dots, m_k\}$$

of $\mathcal{H}$ and complex numbers

$$\lambda_k, \quad k = 1, \dots, K,$$

such that, putting $E_{k,0} := 0$ for simplicity, we have

$$\mathcal{A}E_{k,\ell} = \lambda_k E_{k,\ell} + E_{k,\ell-1}$$

for all k and ℓ. Then the solution of (3.1) is given by the formula

$$U(t) = \sum_{k=1}^{K} \sum_{\ell=1}^{m_k} U_{k,\ell} F_{k,\ell}(t) \quad \text{with} \quad F_{k,\ell}(t) := \sum_{j=0}^{\ell-1} \frac{t^j e^{\lambda_k t}}{j!} E_{k,\ell-j},$$

where the complex numbers $U_{k,\ell}$ are the coefficients of the initial value U_0 in our basis:

$$U_0 = \sum_{k=1}^{K} \sum_{\ell=1}^{m_k} U_{k,\ell} E_{k,\ell}.$$

Indeed, this follows easily from the relations

$$F'_{k,\ell}(t) = \lambda_k F_{k,\ell}(t) + F_{k,\ell-1}(t),$$
$$AF_{k,\ell}(t) = \lambda_k F_{k,\ell}(t) + F_{k,\ell-1}(t),$$

and

$$F_{k,\ell}(0) = E_{k,\ell},$$

where we have put $F_{k,0}(t) := 0$.

Note that since $\mathcal{H}$ is finite-dimensional, we have the estimates

$$c_1 \sum_{k=1}^{K} \sum_{\ell=1}^{m_k} |U_{k,\ell}|^2 \leq \|U_0\|^2 \leq c_2 \sum_{k=1}^{K} \sum_{\ell=1}^{m_k} |U_{k,\ell}|^2$$

with two positive[1] constants c_1, c_2, independent of the particular choice of $U_0 \in \mathcal{H}$.

Remark. In the sequel we often write $A \asymp B$ instead of $c_1 A \leq B \leq c_2 A$ for brevity if we do not need to use explicitly the constants c_1, c_2. For example, the above relation is equivalent to

$$\|U_0\|^2 \asymp \sum_{k=1}^{K} \sum_{\ell=1}^{m_k} |U_{k,\ell}|^2.$$

In all such estimates, the constants c_1, c_2 will be assumed to be independent of the particular choice of the initial data.

Now consider the infinite-dimensional case. The following assumption will be satisfied in almost all examples of this book:

(RB) There exist a *Riesz basis*

$$\{E_{k,\ell} \ : \ k \in K, \quad \ell = 1, \ldots, m_k\}$$

of $\mathcal{H}$ and complex numbers

---

[1] In this book *positive* means *strictly positive*; otherwise, we use the adjective *nonnegative*.

$$\lambda_k, \quad k \in K,$$

such that, putting $E_{k,0} := 0$ for simplicity, we have

$$\mathcal{A} E_{k,\ell} = \lambda_k E_{k,\ell} + E_{k,\ell-1}$$

for all k, ℓ. Moreover,

- $|\lambda_k| \to \infty$, i.e., the family $\{\lambda_k\}$ has no finite accumulation points;
- $\sup |\Re \lambda_k| < \infty$;
- $\sup m_k < \infty$.

In this book, K will always be a *countable* infinite set. Usually K is a set of integers, but sometimes it will be more natural to choose other index sets K.

Let us recall the definition of the Riesz basis: every $U_0 \in \mathcal{H}$ has a unique convergent expansion

$$U_0 = \sum_{k \in K} \sum_{\ell=1}^{m_k} U_{k,\ell} E_{k,\ell}, \tag{3.2}$$

and the coefficients of this expansion satisfy the estimates

$$\|U_0\|^2 \asymp \sum_{k \in K} \sum_{\ell=1}^{m_k} |U_{k,\ell}|^2,$$

i.e.,

$$c_1 \sum_{k \in K} \sum_{\ell=1}^{m_k} |U_{k,\ell}|^2 \le \|U_0\|^2 \le c_2 \sum_{k \in K} \sum_{\ell=1}^{m_k} |U_{k,\ell}|^2, \tag{3.3}$$

with two constants $c_1, c_2 > 0$, independent of the particular choice of $U_0 \in \mathcal{H}$.

Remark. There is no ambiguity about the interpretation of the convergence: as in the case of orthogonal series, the series encountered here and later have only countably many nonzero terms, and they converge unconditionally; i.e., we can arrange the terms in a sequence in an arbitrary order.[2]

Example. The simplest case is that of $\mathcal{A}$ a skew-adjoint operator having a compact resolvent. Then $\mathcal{A}$ is diagonalizable. Moreover, $\mathcal{H}$ has an *orthonormal* basis formed by *ordinary* eigenvectors of $\mathcal{A}$; i.e., $m_k = 1$ for all k, and all eigenvalues are *purely imaginary*. Hence Parseval's equality holds: we may take $c_1 = c_2 = 1$ in (3.3). However, this framework is too restrictive for the study of many natural problems considered in this book.

Let us denote by $\mathcal{Z}$ the linear hull of the basis vectors $E_{k,\ell}$; then $\mathcal{Z}$ is a dense subspace of $\mathcal{H}$. If $U_0 \in \mathcal{Z}$, then its expansion (3.2) has only finitely many nonzero terms, so that it is natural to define the corresponding solution of (3.1) by

---

[2] Alternatively, we could have used the elegant but less familiar concept of sums of families as discussed, e.g., in Halmos [46].

$$U(t) = \sum_{k \in K} \sum_{\ell=1}^{m_k} U_{k,\ell} F_{k,\ell}(t) \quad \text{with} \quad F_{k,\ell}(t) := \sum_{j=0}^{\ell-1} \frac{t^j e^{\lambda_k t}}{j!} E_{k,\ell-j}. \tag{3.4}$$

Clearly, $U : \mathbb{R} \to \mathcal{H}$ belongs to the class C^∞.

As a consequence of assumption (RB) we may define a solution of (3.1) by (3.2) and (3.4) for *all $U_0 \in \mathcal{H}$*:

Theorem 3.1. *Assume (RB). Given $U_0 \in \mathcal{H}$ arbitrarily, the series (3.4), where the coefficients $U_{k,\ell}$ are defined by (3.2), converges, locally uniformly with respect to t, to a continuous function $U : \mathbb{R} \to \mathcal{H}$. It will be called the solution of (3.1).*

Moreover, there exist two continuous and positive functions $c_3, c_4 : \mathbb{R} \to \mathbb{R}$ such that

$$c_3(t)\|U_0\|^2 \le \|U(t)\|^2 \le c_4(t)\|U_0\|^2 \tag{3.5}$$

for all $t \in \mathbb{R}$.

Proof. Let us arrange the terms in a sequence indexed by the positive integers and let us introduce the partial sums

$$U_p(t) := \sum_{k=1}^{p} \sum_{\ell=1}^{m_k} U_{k,\ell} F_{k,\ell}(t)$$

of the series (3.4). Put

$$m := \max m_k \quad \text{and} \quad \alpha := \sup|\Re\lambda_k|$$

for brevity. Using the definition of $F_{k,\ell}(t)$, applying the inequality between arithmetic and quadratic means, and finally using the Riesz basis inequality (3.3), we have for all $p < q$ and $t \in \mathbb{R}$ the following estimates:

$$\|U_q(t) - U_p(t)\|^2 = \left\| \sum_{k=p+1}^{q} \sum_{\ell=1}^{m_k} U_{k,\ell} F_{k,\ell}(t) \right\|^2$$

$$= \left\| \sum_{\ell=1}^{m} \sum_{\substack{p < k \le q: \\ m_k \ge \ell}} U_{k,\ell} F_{k,\ell}(t) \right\|^2$$

$$\le m \sum_{\ell=1}^{m} \left\| \sum_{\substack{p < k \le q: \\ m_k \ge \ell}} U_{k,\ell} F_{k,\ell}(t) \right\|^2$$

$$= m \sum_{\ell=1}^{m} \left\| \sum_{\substack{p < k \le q: \\ m_k \ge \ell}} \sum_{j=0}^{\ell-1} \frac{U_{k,\ell} t^j e^{\lambda_k t}}{j!} E_{k,\ell-j} \right\|^2$$

$$\leq mc_2 \sum_{\ell=1}^{m} \sum_{\substack{p<k\leq q:\\ m_k\geq\ell}} \sum_{j=0}^{\ell-1} \left|\frac{U_{k,\ell} t^j e^{\lambda_k t}}{j!}\right|^2$$

$$\leq mc_2 e^{2\alpha|t|} \sum_{j=0}^{m-1} \frac{t^{2j}}{(j!)^2} \sum_{k=p+1}^{q} \sum_{\ell=1}^{m_k} |U_{k,\ell}|^2.$$

Putting

$$c(t) := mc_2 e^{2\alpha|t|} \sum_{j=0}^{m-1} \frac{t^{2j}}{(j!)^2}$$

we therefore have

$$\|U_q(t) - U_p(t)\|^2 \leq c(t) \sum_{k=p+1}^{q} \sum_{\ell=1}^{m_k} |U_{k,\ell}|^2.$$

The last expression tends to zero as $p \to \infty$, because

$$\sum_{k=1}^{\infty} \sum_{\ell=1}^{m_k} |U_{k,\ell}|^2 \leq \frac{1}{c_1} \|U_0\|^2 < \infty.$$

Furthermore, the convergence is uniform on every bounded interval I because the continuous function $c(t)$ is bounded on I. It follows that $(U_p(t))$ is a Cauchy sequence in $\mathcal{H}$ for each t, and that the Cauchy property is uniform on every bounded interval I. Since $\mathcal{H}$ is complete, U_p converges locally uniformly to a continuous function U.

It remains to establish (3.5). Choosing $p = 0$ in the above computation and then letting $q \to \infty$, we obtain that

$$\|U(t)\|^2 \leq c(t) \sum_{k=1}^{\infty} \sum_{\ell=1}^{m_k} |U_{k,\ell}|^2.$$

Using the Riesz basis property (3.3), we see that the right estimate of (3.5) follows with $c_4(t) = c(t)/c_1$.

The left estimate then easily follows by using the time invariance of the equation (3.1). Indeed, for any fixed $t \in \mathbb{R}$, changing the initial data U_0 to $V_0 := U(t)$ and denoting the corresponding solution of (3.1) by V, we have

$$\|U_0\|^2 = \|V(-t)\|^2 \leq c_4(-t)\|V_0\|^2 = c_4(-t)\|U(t)\|^2,$$

so that the first estimate of (3.5) is satisfied with $c_3(t) = 1/c_4(-t)$.

Remarks.

- For each fixed $t \in \mathbb{R}$, the formula $U_0 \mapsto U(t)$ defines an automorphism of $\mathcal{H}$. Denoting it by $e^{t\mathcal{A}}$, one can readily verify that $e^{0\mathcal{A}}$ is the identity map of $\mathcal{H}$ and that $e^{s\mathcal{A}} e^{t\mathcal{A}} = e^{(s+t)\mathcal{A}}$ for all real s and t. The operator $\mathcal{A}$ is said to generate a *group* $(e^{t\mathcal{A}})$ of automorphisms of $\mathcal{H}$.

- In the skew-adjoint case we have $c_1 = c_2 = 1$ in (3.3) and $c_3(t) = c_4(t) = 1$ in (3.5), so that $\mathcal{A}$ generates a group $\left(e^{t\mathcal{A}}\right)$ of *isometries* of $\mathcal{H}$. The above proof also shows that in this case the convergence $U_A(t) \to U(t)$ is *uniform* in $t \in \mathbb{R}$.
- Fix a real number s and denote by $\mathcal{H}^s$ the completion of $\mathcal{Z}$ (the linear hull of the basis vectors $E_{k,\ell}$) with respect to the Euclidean norm defined by the formula

$$\|U_0\|_{\mathcal{H}^s}^2 = \sum_{k \in K} \left(1 + |\lambda_k|\right)^{2s} \sum_{\ell=1}^{m_k} |U_{k,\ell}|^2;$$

we have $\mathcal{H}^0 = \mathcal{H}$, while for $s > r$ we have $\mathcal{H}^s \subset \mathcal{H}^r$ with a dense and continuous inclusion. It is easy to verify that hypothesis (RB) is satisfied in each space $\mathcal{H}^s$ with the same eigenvalues λ_k and with the renormalized family

$$\left\{\left(1 + |\lambda_k|\right)^{-s} E_{k,\ell} \ : \ k \in K, \quad \ell = 1, \ldots, m_k\right\}$$

as a Riesz basis.[3] Hence Theorem 3.1 remains valid in these spaces. For $s > 0$ this leads to important regularity properties: if the initial value U_0 belongs to the smaller space $\mathcal{H}^s$ instead of $\mathcal{H}$, then the solution of (3.1) is more regular. For $s < 0$ we obtain the existence of solutions of (3.1) if the initial data belong only to a larger space $\mathcal{H}^s$ instead of $\mathcal{H}$: we still have a natural solution, but it is less regular than before.

Let us give some examples.

3.2 Wave Equation with Dirichlet Boundary Condition

Consider the problem

$$\begin{cases} u'' - \Delta u + au = 0 & \text{in} \quad \mathbb{R} \times \Omega, \\ u = 0 & \text{on} \quad \mathbb{R} \times \Gamma, \\ u(0) = u_0 \quad \text{and} \quad u'(0) = u_1 & \text{in} \quad \Omega, \end{cases} \tag{3.6}$$

where Ω is a nonempty bounded open set in $\mathbb{R}^N$ with boundary Γ, and a a real number.[4] In order to simplify the exposition we usually assume that Ω is of class C^∞. The *energy* of the solution is defined by

$$E(t) := \frac{1}{2} \int_\Omega |\nabla u(t,x)|^2 + |u'(t,x)|^2 \ dx, \quad t \in \mathbb{R}.$$

---

[3] If $\lambda_k \neq 0$ for all k, then the above definitions may be simplified by changing $1 + |\lambda_k|$ to $|\lambda_k|$.

[4] Here and in all later examples, we could have chosen for example a real-valued function $a \in L^\infty(\Omega)$ instead of a constant: our proofs would adapt to this more general case without difficulty.

Applying the spectral theorem to the Laplacian operator with homogeneous Dirichlet boundary conditions, we see that $L^2(\Omega)$ has an orthonormal basis $e_1, e_2,\ldots,$ formed by eigenfunctions of $-\Delta$, associated with positive eigenvalues γ_k, tending to ∞:

- e_k belongs to $H_0^1(\Omega)$;
- e_k belongs to $C^\infty(\Omega)$ and $-\Delta e_k = \gamma_k e_k$ in Ω;
- $\gamma_k > 0$;
- $\gamma_k \to \infty$.[5]

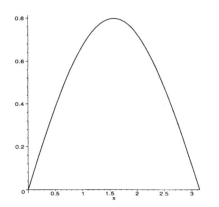

Fig. 3.1. Graph of e_1

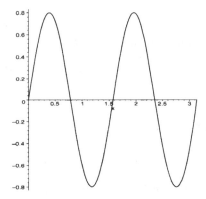

Fig. 3.2. Graph of e_2

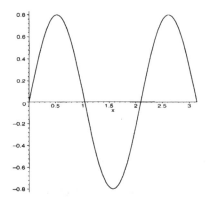

Fig. 3.3. Graph of e_3

Fig. 3.4. Graph of e_4

---

[5]For example, in the one-dimensional case $\Omega = (0, \pi)$ such an orthonormal basis is given by the functions $e_k := \sqrt{2/\pi} \sin kx$, corresponding to the eigenvalues $\gamma_k = k^2$, $k = 1, 2, \ldots$. See Figures 3.1–3.4.

Setting[6]

$$\omega_k := \sqrt{\gamma_k + a},$$

we have the following result:

Proposition 3.2. *If*

$$u_0 \in H_0^1(\Omega) \quad and \quad u_1 \in L^2(\Omega),$$

then (3.6) has a unique solution satisfying

$$u \in C(\mathbb{R}; H_0^1(\Omega)) \cap C^1(\mathbb{R}; L^2(\Omega)).$$

It is given by a series[7]

$$u(t, x) = \sum_{k=1}^{\infty} (a_k e^{i\omega_k t} + a_{-k} e^{-i\omega_k t}) e_k(x)$$

with suitable complex coefficients a_k and a_{-k} such that

$$\sum_{k=1}^{\infty} \gamma_k (|a_k|^2 + |a_{-k}|^2) < \infty.$$

Moreover,

$$E(0) \asymp \sum_{k=1}^{\infty} \gamma_k (|a_k|^2 + |a_{-k}|^2),$$

and there exist two positive functions $c_1, c_2 : \mathbb{R} \to \mathbb{R}$, independent of the choice of the initial data, such that

$$c_1(t)E(0) \le E(t) \le c_2(t)E(0)$$

for all $t \in \mathbb{R}$.

Proof. We rewrite the problem (3.6) in the form (3.1) by setting

$$U := (u, u'), \quad U_0 := (u_0, u_1), \quad and \quad \mathcal{A}(u, v) := (v, \Delta u - au).$$

Let us denote by Z the linear hull of the eigenfunctions e_k and define the domain of $\mathcal{A}$ by $D(\mathcal{A}) := Z \times Z$.

We distinguish three cases.

First case. If $\gamma_k > -a$ for all k, then all numbers ω_k are real and positive. Then one may readily verify that the vectors

---

[6]If $\gamma_k + a < 0$ for some k, then we choose any one of the two square roots. Since $\gamma_k \to \infty$, this can happen only for finitely many indices.

[7]If $\omega_k = 0$ for some k, then the term $a_k e^{i\omega_k t} + a_{-k} e^{-i\omega_k t}$ is replaced by $a_k + a_{-k}t$.

$$E_{\pm k,1} := \frac{1}{\sqrt{\gamma_k + |\omega_k|^2}}(e_k, \pm i\omega_k e_k), \quad k = 1, 2, \ldots,$$

are (ordinary) eigenvectors of $\mathcal{A}$ with purely imaginary eigenvalues $\pm i\omega_k$. Moreover, they form an orthonormal basis in the Hilbert space

$$\mathcal{H} := H_0^1(\Omega) \times L^2(\Omega), \quad \|(u,v)\|_{\mathcal{H}} := \left(\int_\Omega |\nabla u|^2 + |v|^2 \, dx\right)^{1/2}.$$

Hence the assumptions of Theorem 3.1 are satisfied with $m_k = 1$ for all $k \in \mathbb{Z}\backslash\{0\}$: we are in the skew-adjoint case. Putting $\omega_{-k} := -\omega_k$, we see that the corresponding solution of (3.1) is given by the series

$$U(t) = \sum_{k \in \mathbb{Z}\backslash\{0\}} U_{k,1} e^{i\omega_k t} E_{k,1},$$

and its first component provides the expression of the solution $u(t,x)$ of (3.6) with

$$a_{\pm k} = \frac{U_{k,1}}{\sqrt{\gamma_k + |\omega_k|^2}}, \quad k = 1, 2, \ldots.$$

The desired estimates follow by noting that

$$E(t) = \frac{1}{2}\|U(t)\|^2 \quad \text{and} \quad \gamma_k + |\omega_k|^2 \asymp \gamma_k.$$

Second case. If some of the numbers $\gamma_k + a$ are negative but none of them is equal to zero, then the above defined ordinary eigenvectors still form an orthonormal basis of $\mathcal{H}$, but a finite number of the eigenvalues $\pm i\omega_k$ are no longer purely imaginary. Otherwise, the proof remains unchanged.

Third case. If $\omega_k = 0$ for some k (this can happen only for finitely many indices), then we replace $E_{k,1}$ and $E_{-k,1}$ by

$$E_{k,1} := (e_k, 0) \quad \text{and} \quad E_{k,2} := (0, e_k)$$

in the proof. Then

$$U_{k,1} e^{i\omega_k t} E_{k,1} + U_{-k,1} e^{-i\omega_k t} E_{-k,1}$$

in the expression of $U(t)$ is replaced by

$$U_{k,1} E_{k,1} + U_{k,2}\left(E_{k,2} + t E_{k,1}\right),$$

which provides the desired expression of $u(t,x)$ with

$$a_k = U_{k,1} \quad \text{and} \quad a_{-k} = U_{k,2}.$$

Remark. Using the last remark of the preceding section, we may obtain an infinite family of variants of Proposition 3.2 by strengthening or weakening the norms and the "energy" of the solutions. Indeed, fix a real number s and denote by D^s the completion of Z with respect to the Euclidean norm defined by the formula

$$\left\| \sum_{k=1}^{\infty} a_k e_k \right\|_s^2 := \sum_{k=1}^{\infty} (1 + \gamma_k)^s |a_k|^2.$$

For some particular values of s, D^s is a usual Sobolev space; for example,

$$D^0 = L^2(\Omega), \quad D^1 = H_0^1(\Omega), \quad D^2 = H^2(\Omega) \cap H_0^1(\Omega), \quad D^{-1} = H^{-1}(\Omega).$$

One may readily verify that $\mathcal{H}^s = D^{s+1} \times D^s$ (up to a norm equivalence) with the notation of the preceding proof. This leads to a generalization of the preceding proposition, by replacing

$$H_0^1(\Omega), \quad L^2(\Omega), \quad \gamma_k, \quad \text{and} \quad E(t)$$

by

$$D^{s+1}, \quad D^s, \quad \gamma_k^{s+1}, \quad \text{and} \quad E_s(t) := \frac{1}{2}\left(\|u(t)\|_{s+1}^2 + \|u'(t)\|_s^2 \right),$$

respectively, where $u(t, x)$ is still given by the same formula.

3.3 Wave Equation with Neumann Boundary Condition

Next consider the problem

$$\begin{cases} u'' - \Delta u + au = 0 & \text{in} \quad \mathbb{R} \times \Omega, \\ \partial_\nu u = 0 & \text{on} \quad \mathbb{R} \times \Gamma, \\ u(0) = u_0 \quad \text{and} \quad u'(0) = u_1 & \text{in} \quad \Omega, \end{cases} \tag{3.7}$$

with Ω, Γ, and a as in the preceding section. Here and in the sequel we denote by ν the outward unit normal vector to Γ, and by $\frac{\partial u}{\partial \nu}$ or $\partial_\nu u$ the normal derivative of u. For convenience, the "energy" of the solution is defined by

$$E(t) := \frac{1}{2} \int_\Omega |u(t, x)|^2 + |\nabla u(t, x)|^2 + |u'(t, x)|^2 \, dx, \quad t \in \mathbb{R}.$$

It follows from the spectral theorem applied to the Laplacian operator with homogeneous Neumann boundary condition that $L^2(\Omega)$ has an orthonormal basis $e_1, e_2, \ldots$, formed by eigenfunctions of $-\Delta$, associated with nonnegative eigenvalues γ_k, tending to ∞, only one of which is equal to zero:

- e_k belongs to $H^2(\Omega)$ and $\partial_\nu e_k = 0$ on Γ;
- e_k belongs to $C^\infty(\Omega)$ and $-\Delta e_k = \gamma_k e_k$ in Ω;
- $\gamma_1 = 0$ and $\gamma_k > 0$ if $k \geq 2$;

- $\gamma_k \to \infty.$[8]

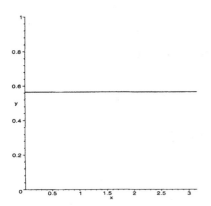

Fig. 3.5. Graph of e_1

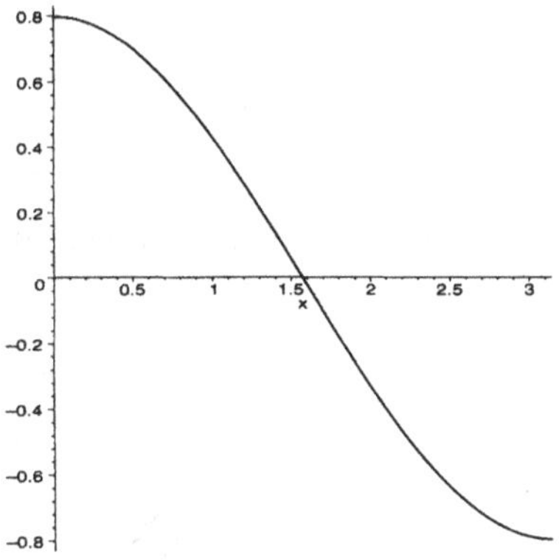

Fig. 3.6. Graph of e_2

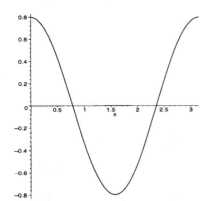
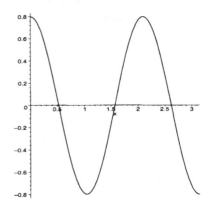

Fig. 3.7. Graph of e_3

Fig. 3.8. Graph of e_4

Setting[9]

$$\omega_k := \sqrt{\gamma_k + a}$$

we have the following proposition:

Proposition 3.3. *If*

$$u_0 \in H^1(\Omega) \quad and \quad u_1 \in L^2(\Omega),$$

then (3.7) has a unique solution satisfying

---

[8]For example, in the one-dimensional case $\Omega = (0, \pi)$, such an orthonormal basis is given by the functions $e_1 := \sqrt{1/\pi}$ and $e_k := \sqrt{2/\pi} \cos(k - 1)x$ for $k \geq 2$, corresponding to the eigenvalues $\gamma_k = (k - 1)^2$, $k = 1, 2, \ldots$ See Figures 3.5–3.8.

[9]If $\gamma_k + a < 0$ for some k, then we choose any one of the two square roots. As in the preceding section, $\gamma_k + a > 0$ for all but finitely many indices.

$$u \in C(\mathbb{R}; H^1(\Omega)) \cap C^1(\mathbb{R}; L^2(\Omega)).$$

It is given by a series[10]

$$u(t,x) = \sum_{k=1}^{\infty} (a_k e^{i\omega_k t} + a_{-k} e^{-i\omega_k t}) e_k(x)$$

with suitable complex coefficients a_k and a_{-k} such that

$$\sum_{k=1}^{\infty} (1 + \gamma_k)(|a_k|^2 + |a_{-k}|^2) < \infty.$$

Moreover,

$$E(0) \asymp \sum_{k=1}^{\infty} (1 + \gamma_k)(|a_k|^2 + |a_{-k}|^2),$$

and there exist two positive functions $c_1, c_2 : \mathbb{R} \to \mathbb{R}$, independent of the choice of the initial data, such that

$$c_1(t)E(0) \le E(t) \le c_2(t)E(0)$$

for all $t \in \mathbb{R}$.

Proof. We rewrite the problem (3.7) in the form (3.1) by again setting

$$U := (u, u'), \quad U_0 := (u_0, u_1), \quad \text{and} \quad \mathcal{A}(u, v) := (v, \Delta u - au).$$

Let us denote by Z the linear hull of the eigenfunctions e_k and define the domain of $\mathcal{A}$ by $D(\mathcal{A}) := Z \times Z$.

We may repeat the proof of Proposition 3.2 by working in the Hilbert space

$$\mathcal{H} := H^1(\Omega) \times L^2(\Omega), \quad \|(u,v)\| := \left(\int_\Omega |u|^2 + |\nabla u|^2 + |v|^2 \, dx \right)^{1/2},$$

and by changing γ_k to $1 + \gamma_k$ everywhere.

Remark. As in the preceding section, it is possible to formulate an infinite family of analogous propositions by considering more or less regular initial data. Introducing the spaces D^s as in the preceding section but by using the new orthonormal basis (e_k) considered above, we have, for example,

$$D^0 = L^2(\Omega), \quad D^1 = H^1(\Omega), \quad D^2 = \{v \in H^2(\Omega) \ : \ \partial_\nu v = 0 \quad \text{on} \quad \Gamma\}.$$

Now for each fixed real number s a variant of Proposition 3.3 is obtained if we replace

---

[10]If $\omega_k = 0$ for some k, then the term $a_k e^{i\omega_k t} + a_{-k} e^{-i\omega_k t}$ is replaced by $a_k + a_{-k} t$.

$$H^1(\Omega), \quad L^2(\Omega), \quad 1 + \gamma_k, \quad \text{and} \quad E(t),$$

respectively, by

$$D^{s+1}, \quad D^s, \quad (1 + \gamma_k)^{s+1}, \quad \text{and} \quad E_s(t) := \frac{1}{2}(\|u(t)\|^2_{s+1} + \|u'(t)\|^2_s),$$

and we keep the same formula for $u(t, x)$.

We shall use later the case $s = -1$: then the energy is given by the formula

$$E_{-1}(t) := \frac{1}{2} \int_\Omega |u(t,x)|^2 + |\nabla \Delta^{-1} u'(t,x)|^2 \, dx.$$

Here Δ maps D^1 onto D^{-1}, and $\Delta^{-1} v$ is defined only up to an additive constant (a multiple of e_1). But $\nabla \Delta^{-1} v$ is uniquely defined.

3.4 Wave Equation with Mixed Boundary Conditions

Next consider the more complex problem

$$\begin{cases} u'' - \Delta u + au = 0 & \text{in} \quad \mathbb{R} \times \Omega, \\ u = 0 & \text{on} \quad \mathbb{R} \times \Gamma_0, \\ \partial_\nu u = 0 & \text{on} \quad \mathbb{R} \times \Gamma_1, \\ u(0) = u_0 \quad \text{and} \quad u'(0) = u_1 & \text{in} \quad \Omega, \end{cases} \tag{3.8}$$

with the same assumptions and notation as in the preceding two sections. Furthermore, we assume that Γ_0 is a *nonempty* both open and closed subset of Γ, and $\Gamma_1 = \Gamma \backslash \Gamma_0$. If $\Gamma_0 = \Gamma$, then this model reduces to the case of Dirichlet boundary conditions considered in Section 3.2.[11]

Examples.

- The last condition is obviously satisfied in the Dirichlet case in which $\Gamma_0 = \Gamma$ and $\Gamma_1 = \emptyset$.
- The condition is also satisfied if Ω is an annular region and Γ_0, Γ_1 are its outer and inner boundaries, respectively.
- A third case in which this condition is satisfied is that of Ω a one-dimensional interval $(0, \ell)$ and $\Gamma_0 = \{0\}$, $\Gamma_1 = \{\ell\}$.

Under these assumptions the spectral theorem implies the existence of an orthonormal basis $e_1, e_2,\ldots$ of $L^2(\Omega)$, formed by eigenfunctions of $-\Delta$, associated with positive eigenvalues γ_k, tending to ∞:

- e_k belongs to $H^2(\Omega)$, $e_k = 0$ on Γ_0 and $\partial_\nu e_k = 0$ on Γ_1;

---

[11] The above assumptions are satisfied, for example, if Γ_0 is a connected component of the boundary. More general partitions of the boundary were considered, e.g., by Grisvard [45], Komornik and Zuazua [83], and Bey, Lohéac, and Moussaoui [15].

- e_k belongs to $C^\infty(\Omega)$ and $-\Delta e_k = \gamma_k e_k$ in Ω;
- $\gamma_k > 0$ for all k,
- $\gamma_k \to \infty$.[12]

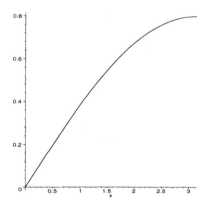

Fig. 3.9. Graph of e_1

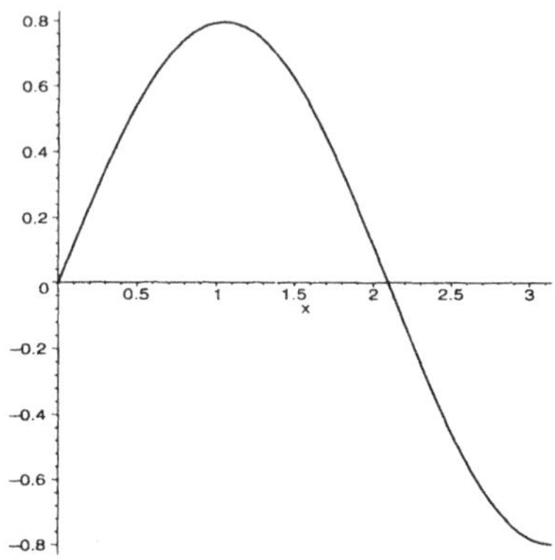

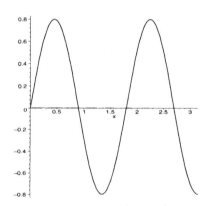

Fig. 3.10. Graph of e_2

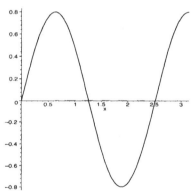

Fig. 3.11. Graph of e_3

Fig. 3.12. Graph of e_4

Let us introduce the Hilbert space

$$H^1_{\Gamma_0}(\Omega) := \{v \in H^1(\Omega) \ : \ v = 0 \quad \text{on} \quad \Gamma_0\}, \quad \|v\| := \left(\int_\Omega |\nabla v|^2 \ dx\right)^{1/2}.$$

---

[12]For example, in the one-dimensional case $\Omega = (0, \pi)$ with $\Gamma_0 = \{0\}$ such an orthonormal basis is given by the functions $e_k := \sqrt{2/\pi} \sin(k - (1/2))x$ for $k \geq 1$, corresponding to the eigenvalues $\gamma_k = (k - (1/2))^2$. See Figures 3.9–3.12.

Setting[13]

$$\omega_k := \sqrt{\gamma_k + a}$$

and defining the "energy" of the solution by the formula

$$E(t) := \frac{1}{2} \int_\Omega |\nabla u(t, x)|^2 + |u'(t, x)|^2 \, dx, \quad t \in \mathbb{R},$$

we have the following result:

Proposition 3.4. *If*

$$u_0 \in H^1_{\Gamma_0}(\Omega) \quad and \quad u_1 \in L^2(\Omega),$$

then (3.8) has a unique solution satisfying

$$u \in C\big(\mathbb{R}; H^1_{\Gamma_0}(\Omega)\big) \cap C^1\big(\mathbb{R}; L^2(\Omega)\big).$$

It is given by a series[14]

$$u(t, x) = \sum_{k=1}^\infty (a_k e^{i\omega_k t} + a_{-k} e^{-i\omega_k t}) e_k(x)$$

with suitable complex coefficients a_k and a_{-k} such that

$$\sum_{k=1}^\infty \gamma_k \big(|a_k|^2 + |a_{-k}|^2\big) < \infty.$$

Moreover,

$$E(0) \asymp \sum_{k=1}^\infty \gamma_k \big(|a_k|^2 + |a_{-k}|^2\big),$$

and there exist two positive functions $c_1, c_2 : \mathbb{R} \to \mathbb{R}$, independent of the choice of the initial data, such that

$$c_1(t) E(0) \le E(t) \le c_2(t) E(0)$$

for all $t \in \mathbb{R}$.

Proof. This is a straightforward generalization of the proof of Proposition 3.2 obtained by working in the Hilbert space

$$\mathcal{H} := H^1_{\Gamma_0}(\Omega) \times L^2(\Omega), \quad \|(u, v)\|_{\mathcal{H}} := \left(\int_\Omega |\nabla u|^2 + |v|^2 \, dx \right)^{1/2}.$$

---

[13]If $\gamma_k + a < 0$ for some k, then we choose any one of the two square roots. As in the preceding section, $\gamma_k + a > 0$ for all but finitely many indices.
[14]If $\omega_k = 0$ for some k, then the term $a_k e^{i\omega_k t} + a_{-k} e^{-i\omega_k t}$ is replaced by $a_k + a_{-k} t$.

3.5 A Petrovsky System with Hinged Boundary Conditions

Now consider the problem

$$\begin{cases} u'' + \Delta^2 u + au = 0 & \text{in } \mathbb{R} \times \Omega, \\ u = \Delta u = 0 & \text{on } \mathbb{R} \times \Gamma, \\ u(0) = u_0 \quad \text{and} \quad u'(0) = u_1 & \text{in } \Omega. \end{cases} \tag{3.9}$$

Define the energy of the solutions by the formula

$$E(t) := \frac{1}{2} \int_\Omega |u(t,x)|^2 + |\Delta u(t,x)|^2 + |u'(t,x)|^2 \, dx, \quad t \in \mathbb{R}.$$

Consider in the Hilbert space $L^2(\Omega)$ the same orthonormal basis e_1, $e_2, \ldots$ as in the case of the wave equation with Dirichlet boundary condition in Section 3.2, and set[15]

$$\omega_k := \sqrt{\gamma_k^2 + a}.$$

Proposition 3.5. *If*

$$u_0 \in H^2(\Omega) \cap H_0^1(\Omega) \qquad \text{and} \qquad u_1 \in L^2(\Omega),$$

then (3.9) has a unique solution

$$u \in C(\mathbb{R}; H^2(\Omega) \cap H_0^1(\Omega)) \cap C^1(\mathbb{R}; L^2(\Omega)).$$

It is given by a series[16]

$$u(t,x) = \sum_{k=1}^\infty (a_k e^{i\omega_k t} + a_{-k} e^{-i\omega_k t}) e_k(x)$$

with suitable complex coefficients a_k and a_{-k} such that

$$\sum_{k=1}^\infty \gamma_k^2 \big(|a_k|^2 + |a_{-k}|^2\big) < \infty.$$

Moreover,

$$E(0) \asymp \sum_{k=1}^\infty \gamma_k^2 \big(|a_k|^2 + |a_{-k}|^2\big),$$

and there exist two positive functions $c_1, c_2 : \mathbb{R} \to \mathbb{R}$, independent of the choice of the initial data, such that

$$c_1(t)E(0) \leq E(t) \leq c_2(t)E(0)$$

for all $t \in \mathbb{R}$.

---

[15] If $\gamma_k^2 + a < 0$ for some k, then we choose any one of the two square roots.
[16] If $\omega_k = 0$ for some k, then the term $a_k e^{i\omega_k t} + a_{-k} e^{-i\omega_k t}$ is replaced by $a_k + a_{-k} t$.

Proof. We rewrite the problem (3.9) in the form (3.1) by setting

$$U := (u, u'), \quad U_0 := (u_0, u_1), \quad \text{and} \quad \mathcal{A}(u, v) := (v, -\Delta^2 u - au).$$

Let us denote by Z the linear hull of the eigenfunctions e_k and define the domain of $\mathcal{A}$ by $D(\mathcal{A}) := Z \times Z$. We may repeat the proof of Proposition 3.2 by working this time in the Hilbert space

$$\mathcal{H} := \left(H^2(\Omega) \cap H_0^1(\Omega)\right) \times L^2(\Omega), \quad \|(u, v)\|_{\mathcal{H}} := \left(\int_\Omega |u|^2 + |\Delta u|^2 + |v|^2 \, dx\right)^{1/2},$$

and by changing γ_k to γ_k^2 everywhere in the proof.

As in the preceding problems, we may obtain infinitely many variants of the above proposition. Indeed, introducing the *same* Hilbert spaces D^s as in Section 3.2 (p. 38), we may replace for every fixed real number s,

$$H^2(\Omega) \cap H_0^1(\Omega), \quad L^2(\Omega), \quad \gamma_k^2, \quad \text{and} \quad E(t),$$

respectively, by

$$D^{s+2}, \quad D^s, \quad \gamma_k^{s+2}, \quad \text{and} \quad E_s(t) := \frac{1}{2}\left(\|u(t, \cdot)\|_{s+2}^2 + \|u'(t, \cdot)\|_s^2\right).$$

Let us mention two particular cases for later reference:

- (Case $s = 1$.) For

$$u_0 \in D^3 = \{v \in H^3(\Omega) \ : \ v = \Delta v = 0 \quad \text{on} \quad \Gamma\} \quad \text{and} \quad u_1 \in H_0^1(\Omega)$$

the solutions of (3.9) satisfy the estimates

$$E_1(0) \asymp \sum_{k=1}^{\infty} \gamma_k^3 \left(|a_k|^2 + |a_{-k}|^2\right),$$

where the solution is given by the same series as before, and

$$E_1(t) := \frac{1}{2} \int_\Omega |\nabla \Delta u(t, x)|^2 + |\nabla u'(t, x)|^2 \, dx, \quad t \in \mathbb{R}.$$

Indeed, we have only to observe that the formula

$$\|v\| := \frac{1}{2} \int_\Omega |\nabla \Delta v|^2 \, dx$$

defines an equivalent norm on D^3.

- (Case $s = -1$.) For $u_0 \in H_0^1(\Omega)$ and $u_1 \in H^{-1}(\Omega)$ the solutions of (3.9) satisfy the estimates

$$E_{-1}(0) \asymp \sum_{k=1}^{\infty} \gamma_k \left(|a_k|^2 + |a_{-k}|^2 \right),$$

where the solution is given by the same series as before, and

$$E_{-1}(t) := \frac{1}{2} \int_{\Omega} |\nabla u(t,x)|^2 + |\nabla \Delta^{-1} u'(t,x)|^2 \, dx, \quad t \in \mathbb{R},$$

where Δ^{-1} denotes the inverse of the *isometric anti-isomorphism* $\Delta : H_0^1(\Omega) \to H^{-1}(\Omega)$.

3.6 A Petrovsky System with Guided Boundary Conditions

Next consider the problem

$$\begin{cases} u'' + \Delta^2 u + au = 0 & \text{in} \quad \mathbb{R} \times \Omega, \\ \partial_\nu u = \partial_\nu \Delta u = 0 & \text{on} \quad \mathbb{R} \times \Gamma, \\ u(0) = u_0 \quad \text{and} \quad u'(0) = u_1 & \text{in} \quad \Omega. \end{cases} \tag{3.10}$$

Define the energy of the solution by

$$E(t) := \frac{1}{2} \int_{\Omega} |u(t,x)|^2 + |\Delta u(t,x)|^2 + |u'(t,x)|^2 \, dx, \quad t \in \mathbb{R}.$$

Consider in the Hilbert space $H := L^2(\Omega)$ the same orthonormal basis e_1, $e_2, \ldots$ as in the case of the wave equation with Neumann boundary condition in Section 3.3, and set[17]

$$\omega_k := \sqrt{\gamma_k^2 + a}.$$

Introducing the Hilbert space

$$V := \{v \in H^2(\Omega) \ : \ \partial_\nu v = 0 \quad \text{on} \quad \Gamma\}$$

with the norm

$$\|v\|_V := \left(\int_{\Omega} |v|^2 + |\Delta v|^2 \, dx \right)^{1/2},$$

we have the following proposition:

---

[17]If $\gamma_k^2 + a < 0$ for some k, then we choose any one of the two square roots.

Proposition 3.6. *If $u_0 \in V$ and $u_1 \in H$, then (3.10) has a unique solution*

$$u \in C(\mathbb{R}; V) \cap C^1(\mathbb{R}; H).$$

It is given by a series[18]

$$u(t, x) = \sum_{k=1}^{\infty} (a_k e^{i\omega_k t} + a_{-k} e^{-i\omega_k t}) e_k(x)$$

with suitable complex coefficients a_k and a_{-k} such that

$$\sum_{k=1}^{\infty} (1 + \gamma_k)^2 \left(|a_k|^2 + |a_{-k}|^2 \right) < \infty.$$

Moreover,

$$E(0) \asymp \sum_{k=1}^{\infty} (1 + \gamma_k)^2 \left(|a_k|^2 + |a_{-k}|^2 \right),$$

and there exist two positive functions $c_1, c_2 : \mathbb{R} \to \mathbb{R}$, independent of the choice of the initial data, such that

$$c_1(t) E(0) \le E(t) \le c_2(t) E(0)$$

for all $t \in \mathbb{R}$.

Proof. We rewrite the problem (3.10) in the form (3.1) by setting

$$U := (u, u'), \quad U_0 := (u_0, u_1), \quad \text{and} \quad \mathcal{A}(u, v) := (v, -\Delta^2 u - au).$$

Let us denote by Z the linear hull of the eigenfunctions e_k and define the domain of $\mathcal{A}$ by $D(\mathcal{A}) := Z \times Z$. Then we can repeat the proof of Proposition 3.2, working this time in the Hilbert space $\mathcal{H} := V \times H$.

As before, we obtain many variants of the above proposition by introducing the *same* Hilbert spaces D^s as in Section 3.3 (p. 42), and replacing for any fixed real number s,

$$V, \quad H, \quad (1 + \gamma_k)^2, \quad \text{and} \quad E(t),$$

respectively, by

$$D^{s+2}, \quad D^s, \quad (1 + \gamma_k)^{s+2}, \quad \text{and} \quad E_s(t) := \frac{1}{2} \left(\|u(t, \cdot)\|_{s+2}^2 + \|u'(t, \cdot)\|_s^2 \right).$$

We shall use later the case $s = -2$: then the energy is given by

$$E_{-2}(t) := \frac{1}{2} \int_{\Omega} |u(t, x)|^2 + |\Delta^{-1} u'(t, x)|^2 \, dx, \quad t \in \mathbb{R}.$$

---

[18] If $\omega_k = 0$ for some k, then the term $a_k e^{i\omega_k t} + a_{-k} e^{-i\omega_k t}$ is replaced by $a_k + a_{-k} t$.

3.7 A Petrovsky System with Mixed Boundary Conditions

Consider finally the problem

$$\begin{cases} u'' + \Delta^2 u + au = 0 & \text{in} \quad \mathbb{R} \times \Omega, \\ u = \Delta u = 0 & \text{on} \quad \mathbb{R} \times \Gamma_0, \\ \partial_\nu u = \partial_\nu \Delta u = 0 & \text{on} \quad \mathbb{R} \times \Gamma_1, \\ u(0) = u_0 \quad \text{and} \quad u'(0) = u_1 & \text{in} \quad \Omega, \end{cases} \qquad (3.11)$$

with the same assumptions and notation as in Section 3.4. Let us also introduce the same orthonormal basis $e_1, e_2, \ldots$ of $H := L^2(\Omega)$ and the same eigenvalues γ_k.

Define the energy of the solution by

$$E(t) := \frac{1}{2} \int_\Omega |u(t,x)|^2 + |\Delta u(t,x)|^2 + |u'(t,x)|^2 \, dx, \quad t \in \mathbb{R}.$$

Introducing the Hilbert space

$$V := \{ v \in H^2(\Omega) \ : \ v = 0 \quad \text{on} \quad \Gamma_0 \quad \text{and} \quad \partial_\nu v = 0 \quad \text{on} \quad \Gamma_1 \}$$

with the norm

$$\|v\|_V := \left(\int_\Omega |v|^2 + |\Delta v|^2 \, dx \right)^{1/2}$$

and setting

$$\omega_k := \sqrt{\gamma^2 + a},$$

we have the following generalization of Proposition 3.5 (p. 48):

Proposition 3.7. *If $u_0 \in V$ and $u_1 \in H$, then (3.11) has a unique solution*

$$u \in C(\mathbb{R}; V) \cap C^1(\mathbb{R}; H).$$

It is given by a series[19]

$$u(t,x) = \sum_{k=1}^\infty (a_k e^{i\omega_k t} + a_{-k} e^{-i\omega_k t}) e_k(x)$$

with suitable complex coefficients a_k and a_{-k} such that

$$\sum_{k=1}^\infty (1 + \gamma_k)^2 \left(|a_k|^2 + |a_{-k}|^2 \right) < \infty.$$

Moreover,

---

[19]If $\omega_k = 0$ for some k, then the term $a_k e^{i\omega_k t} + a_{-k} e^{-i\omega_k t}$ is replaced by $a_k + a_{-k} t$.

$$E(0) \asymp \sum_{k=1}^{\infty} (1+\gamma_k)^2 \left(|a_k|^2 + |a_{-k}|^2 \right),$$

and there exist two positive functions $c_1, c_2 : \mathbb{R} \to \mathbb{R}$, independent of the choice of the initial data, such that

$$c_1(t)E(0) \le E(t) \le c_2(t)E(0)$$

for all $t \in \mathbb{R}$.

Proof. We rewrite the problem (3.11) in the form (3.1) by setting

$$U := (u, u'), \quad U_0 := (u_0, u_1), \quad \text{and} \quad \mathcal{A}(u, v) := (v, -\Delta^2 u - au).$$

Let us denote by Z the linear hull of the eigenfunctions e_k and define the domain of $\mathcal{A}$ by $D(\mathcal{A}) := Z \times Z$. Then we can repeat the proof of Proposition 3.2, working this time in the Hilbert space $\mathcal{H} := V \times H$.

3.8 A Coupled System

Fix four complex numbers a, b, c, d and consider the following coupled system formed by a wave equation and a Petrovsky system:

$$
\begin{cases}
u'' - \Delta u + au + bw = 0 & \text{in } \mathbb{R} \times \Omega, \\
w'' + \Delta^2 w + cu + dw = 0 & \text{in } \mathbb{R} \times \Omega, \\
u = 0 & \text{on } \mathbb{R} \times \Gamma, \\
w = \Delta w = 0 & \text{on } \mathbb{R} \times \Gamma, \\
u(0) = u_0 \quad \text{and} \quad u'(0) = u_1 & \text{in } \Omega, \\
w(0) = w_0 \quad \text{and} \quad w'(0) = w_1 & \text{in } \Omega.
\end{cases}
\tag{3.12}
$$

We define the *energy* of the solution by the formula[20]

$$E(t) := \frac{1}{2} \int_{\Omega} |\nabla u(t,x)|^2 + |u'(t,x)|^2 + |\nabla w(t,x)|^2 + |\nabla \Delta^{-1} w'(t,x)|^2 \, dx, \quad t \in \mathbb{R}.$$

The well-posedness of this problem easily follows from the earlier results if we appply the basic theory of semigroups:

Proposition 3.8. *Given*

$$u_0 \in H_0^1(\Omega), \quad u_1 \in L^2(\Omega), \quad w_0 \in H_0^1(\Omega), \quad \text{and} \quad w_1 \in H^{-1}(\Omega)$$

arbitrarily, the problem (3.12) has a unique solution satisfying

$$u \in C(\mathbb{R}; H_0^1(\Omega)) \cap C^1(\mathbb{R}; L^2(\Omega))$$

and

$$w \in C(\mathbb{R}; H_0^1(\Omega)) \cap C^1(\mathbb{R}; H^{-1}(\Omega)).$$

---

[20] As in Section 3.5, Δ^{-1} denotes the inverse of the restriction of Δ to $H_0^1(\Omega)$.

Proof. If $b = c = 0$, then the system is uncoupled, and the result follows at once from Proposition 3.2 and the variant $s = -1$ of Proposition 3.5 (pp. 40 and 48). The general case then easily follows from the theory of semigroups, because the terms bw and cu represent a *bounded perturbation* of the infinitesimal generator of the semigroup associated with the uncoupled system.

Let us also give a direct proof by applying the abstract Theorem 3.1. For this, let us rewrite the problem (3.12) in the form

$$U' = \mathcal{A}U, \quad U(0) = U_0,$$

by putting

$$U = (u, w, u', w'), \quad U_0 = (u_0, w_0, u_1, w_1),$$

and

$$\mathcal{A}(u, w, v, z) := (v, z, \Delta u - au - bw, -\Delta^2 w - cu - dw)$$

with

$$D(\mathcal{A}) := Z \times Z \times Z \times Z,$$

where Z denotes the linear hull of the Dirichlet eigenfunctions $e_1, e_2, \ldots$ as in Sections 3.2 and 3.5 (pp. 38, 48), with the same eigenvalues $\gamma_1, \gamma_2, \ldots$ as there.

Let us also introduce the Hilbert space

$$\mathcal{H} := H_0^1(\Omega) \times H_0^1(\Omega) \times L^2(\Omega) \times H^{-1}(\Omega);$$

then

$$E(t) = \frac{1}{2}\|U(t)\|^2.$$

Proposition 3.9. *The operator $\mathcal{A}$ satisfies hypothesis (RB) (p. 34) in the Hilbert space $\mathcal{H}$. Moreover, we can choose the corresponding Riesz basis in the form*

$$\{E_{k,j} \ : \ k = 1, 2, \ldots, \quad j = 1, 2, 3, 4\}$$

with corresponding eigenvalues $i\omega_{k,j}$ so as to have the following asymptotic behavior as $k \to \infty$:

$$E_{k,1} = \begin{pmatrix} \gamma_k^{-1/2} \\ o(1) \\ i + o(1) \\ o(1) \end{pmatrix} e_k, \qquad E_{k,2} = \begin{pmatrix} \gamma_k^{-1/2} \\ o(1) \\ -i + o(1) \\ o(1) \end{pmatrix} e_k, \qquad (3.13)$$

$$E_{k,3} = \begin{pmatrix} o(1) \\ \gamma_k^{-1/2} \\ o(1) \\ \gamma_k^{1/2}(i + o(1)) \end{pmatrix} e_k, \qquad E_{k,4} = \begin{pmatrix} o(1) \\ \gamma_k^{-1/2} \\ o(1) \\ \gamma_k^{1/2}(-i + o(1)) \end{pmatrix} e_k, \qquad (3.14)$$

and

$$\omega_{k,1} = \sqrt{\gamma_k} + o(1), \qquad\qquad \omega_{k,2} = -\sqrt{\gamma_k} + o(1), \qquad (3.15)$$
$$\omega_{k,3} = \gamma_k + o(1), \qquad\qquad \omega_{k,4} = -\gamma_k + o(1). \qquad (3.16)$$

Furthermore, we have $\omega_{k,2} = -\omega_{k,1}$ and $\omega_{k,4} = -\omega_{k,3}$ for all k.

Finally, if a, b, c, d are real numbers, then the numbers $\omega_{k,j}$ are also real for all sufficiently large k.

Proof. Consider the 4×4 matrices

$$\mathcal{A}_k := \begin{pmatrix} 0 & 0 & 1 & 0 \\ 0 & 0 & 0 & 1 \\ -\gamma_k - a & -b & 0 & 0 \\ -c & -\gamma_k^2 - d & 0 & 0 \end{pmatrix}, \quad k = 1, 2, \dots.$$

For each fixed k, by Jordan's theorem $\mathbb{C}^4$ has a basis $A_{k,1}$, $A_{k,2}$, $A_{k,3}$, $A_{k,4}$ formed by ordinary or generalized eigenvectors of $\mathcal{A}_k$ with corresponding eigenvalues $i\omega_{k,1}$, $i\omega_{k,2}$, $i\omega_{k,3}$, and $i\omega_{k,4}$ such that $\omega_{k,2} = -\omega_{k,1}$, $\omega_{k,4} = -\omega_{k,3}$, and

$$-2\omega_{k,1}^2 = -2\omega_{k,2}^2 = \gamma_k^2 + d + \gamma_k + a + \sqrt{\left(\gamma_k^2 + d - \gamma_k - a\right)^2 + 4bc}$$

and

$$-2\omega_{k,3}^2 = -2\omega_{k,4}^2 = \gamma_k^2 + d + \gamma_k + a - \sqrt{\left(\gamma_k^2 + d - \gamma_k - a\right)^2 + 4bc}.$$

Since $\gamma_k \to \infty$, the relations (3.15)–(3.16) follow by a direct computation. It follows that $\mathcal{A}_k$ has four distinct eigenvalues if k is sufficiently large. Furthermore, a direct computation also shows that we can normalize the eigenvectors $A_{k,j}$, so that, by putting $E_{k,j} := A_{k,j} e_k$ the relations (3.13)–(3.14) are also satisfied.

It is clear from the construction that the linear hull of the family $\{E_{k,j}\}$ is equal to $Z \times Z \times Z \times Z$, so that it is dense in $\mathcal{H}$. The Riesz basis property now easily follows by observing that the four-dimensional subspaces spanned by $\{E_{k,1}, E_{k,2}, E_{k,3}, E_{k,4}\}$ are mutually orthogonal and that the four spanning vectors closer and closer to being orthogonal as $k \to \infty$; i.e.,

$$\frac{(E_{k,j}, E_{k,n})_{\mathcal{H}}}{\|E_{k,j}\| \cdot \|E_{k,n}\|} \to 0$$

if $j \neq n$.

Analogous results hold if we change the boundary conditions in (3.12) as follows:

$$\begin{cases} u'' - \Delta u + au + bw = 0 & \text{in} \ \ \mathbb{R} \times \Omega, \\ w'' + \Delta^2 w + cu + dw = 0 & \text{in} \ \ \mathbb{R} \times \Omega, \\ \partial_\nu u = 0 & \text{on} \ \ \mathbb{R} \times \Gamma, \\ \partial_\nu w = \partial_\nu \Delta w = 0 & \text{on} \ \ \mathbb{R} \times \Gamma, \\ u(0) = u_0 \ \ \text{and} \ \ u'(0) = u_1 & \text{in} \ \ \Omega, \\ w(0) = w_0 \ \ \text{and} \ \ w'(0) = w_1 & \text{in} \ \ \Omega. \end{cases} \qquad (3.17)$$

Setting

$$V := \{v \in H^2(\Omega) \; : \; \partial_\nu v = 0 \quad \text{on} \quad \Gamma\}$$

and

$$\|v\|_V := \left(\int_\Omega |v|^2 + |\Delta v|^2 \; dx \right)^{1/2}$$

for brevity, we see that the new problem is well-posed in the following sense:

Proposition 3.10. *For arbitrary*

$$u_0 \in H^1(\Omega), \quad u_1 \in L^2(\Omega), \quad w_0 \in V, \quad and \quad w_1 \in L^2(\Omega),$$

the problem (3.17) *has a unique solution satisfying*

$$u \in C(\mathbb{R}; H^1(\Omega)) \cap C^1(\mathbb{R}; L^2(\Omega))$$

and

$$w \in C(\mathbb{R}; V) \cap C^1(\mathbb{R}; L^2(\Omega)).$$

Proof. If $b = c = 0$, then the system is uncoupled, and the result follows at once from Propositions 3.3 and 3.6 (pp. 43 and 51). The general case then easily follows from the theory of semigroups, because the terms bw and cu represent a *bounded perturbation* of the infinitesimal generator of the semigroup associated with the uncoupled system.

Remark. We can rewrite the problem (3.17) in the form

$$U' = \mathcal{A}U, \quad U(0) = U_0,$$

with the only change that now Z denotes the linear hull of the Neumann eigenfunctions $e_1, e_2, \ldots$ as in Sections 3.3 and 3.6 (pp. 42, 50). Introducing now the Hilbert space

$$\mathcal{H} := H^1(\Omega) \times V \times L^2(\Omega) \times L^2(\Omega),$$

we see that Proposition 3.9 remains valid verbatim, with the same proof.

4

Observability of Strings

In this chapter we introduce two important generalizations of Parseval's equality. They enable us to solve several simple but already nontrivial problems concerning the observability of strings.

Throughout this chapter it will be convenient to use the notation $\mu_k := \sqrt{\gamma_k}$; then $1 + \mu_k \asymp k$.

4.1 Strings with Free Endpoints I

We begin by generalizing and completing Proposition 1.1 (p. 2). Fix a positive number ℓ, a real number a, and consider the following system:

$$
\begin{cases}
u_{tt} - u_{xx} + au = 0 & \text{in} \quad \mathbb{R} \times (0, \ell), \\
u_x(t, 0) = u_x(t, \ell) = 0 & \text{for} \quad t \in \mathbb{R}, \\
u(0, x) = u_0(x) & \text{for} \quad x \in (0, \ell), \\
u_t(0, x) = u_1(x) & \text{for} \quad x \in (0, \ell).
\end{cases}
\tag{4.1}
$$

This is a special case of the system considered in Section 3.3 (p. 42), corresponding to $\Omega = (0, \ell)$. Putting

$$\mu_k := (k - 1)\pi/\ell$$

for brevity, we may choose in $L^2(0, \ell)$ the orthonormal basis

$$e_1(x) = \sqrt{1/\ell} \quad \text{and} \quad e_k(x) = \sqrt{2/\ell} \cos \mu_k x, \quad k = 2, 3, \dots.$$

Since $\gamma_k = \mu_k^2$, putting

$$\omega_k := \sqrt{\mu_k^2 + a}$$

and

$$E(t) := \frac{1}{2} \int_0^\ell |u(t, x)|^2 + |u_x(t, x)|^2 + |u_t(t, x)|^2 \ dx,$$

we have the following special case of Proposition 3.3 (p. 43):

Proposition 4.1. *If*

$$u_0 \in H^1(0, \ell) \quad and \quad u_1 \in L^2(0, \ell),$$

then (4.1) has a unique solution satisfying

$$u \in C(\mathbb{R}; H^1(0, \ell)) \cap C^1(\mathbb{R}; L^2(0, \ell)).$$

It is given by a series[1]

$$u(t, x) = \sum_{k=1}^{\infty} (a_k e^{i\omega_k t} + a_{-k} e^{-i\omega_k t}) \cos \mu_k x$$

with suitable complex coefficients a_k and a_{-k} such that

$$\sum_{k=1}^{\infty} k^2 (|a_k|^2 + |a_{-k}|^2) < \infty.$$

Moreover, we have

$$E(0) \asymp \sum_{k=1}^{\infty} k^2 (|a_k|^2 + |a_{-k}|^2),$$

and there exist two strictly positive functions $c_3, c_4 : \mathbb{R} \to \mathbb{R}$, independent of the choice of the initial data, such that

$$c_3(t) E(0) \le E(t) \le c_4(t) E(0)$$

for all $t \in \mathbb{R}$.

In order to formulate our next result concerning the observability at one of the endpoints, let us introduce for every positive integer k' the following finite-codimensional subspaces of the Hilbert spaces $H^1(0, \ell)$ and $L^2(0, \ell)$:

$$H_{k'} := \left\{ v \in L^2(0, \ell) \ : \ \int_0^\ell v(x) \cos \mu_k x \, dx = 0 \quad \text{for all} \quad 0 \le k < k' \right\},$$

$$V_{k'} := \left\{ v \in H^1(0, \ell) \ : \ \int_0^\ell v(x) \cos \mu_k x \, dx = 0 \quad \text{for all} \quad 0 \le k < k' \right\}.$$

Observe that the functions

$$\cos \mu_k x, \quad k = k', k' + 1, \dots,$$

form an orthogonal basis in both $H_{k'}$ and $V_{k'}$.

---

[1] If $\omega_k = 0$ for some k, then the term $a_k e^{i\omega_k t} + a_{-k} e^{-i\omega_k t}$ is replaced by $a_k + a_{-k} t$. Note that this cannot happen for more than one index k.

Proposition 4.2. *Given an interval I of length $|I| > 2\ell$, there exists an integer k' such that we have*

$$\|u(\cdot,0)\|_{H^1(I)}^2 \asymp \|u(\cdot,\ell)\|_{H^1(I)}^2 \asymp E(0)$$

for all solutions of (4.1) corresponding to initial data $u_0 \in V_{k'}$ and $u_1 \in H_{k'}$.

Remark. For $a = 0$ the optimality of the condition $|I| > 2\ell$ can be easily proved by applying Parseval's equality. More precisely, one can show in this way that the estimates fail if $|I| < 2\ell$. For $a \neq 0$ the optimality of the condition $|I| > 2\ell$ follows at once by applying Beurling's Theorem 9.2, to be discussed later (p. 174).

Proof. Fix k' satisfying

$$\omega_k > 0 \quad \text{for all} \quad k \geq k',$$

and sufficiently large as to be specified later.

In view of the preceding proposition we have to establish the following estimates:[2]

$$\int_I |u(t,0)|^2 \, dt + \int_I |u_t(t,0)|^2 \, dt \asymp \sum_{k=k'}^{\infty} k^2 \big(|a_k|^2 + |a_{-k}|^2\big). \tag{4.2}$$

Observe that

$$\int_I |u(t,0)|^2 \, dt = \int_I \Big| \sum_{k=k'}^{\infty} \big(a_k e^{i\omega_k t} + a_{-k} e^{-i\omega_k t}\big) \Big|^2 \, dt \tag{4.3}$$

and

$$\int_I |u_t(t,0)|^2 \, dt = \int_I \Big| \sum_{k=k'}^{\infty} \omega_k \big(a_k e^{i\omega_k t} - a_{-k} e^{-i\omega_k t}\big) \Big|^2 \, dt. \tag{4.4}$$

If $a \neq 0$, then the functions $e^{\pm i\omega_k t}$ are *not orthogonal* on any interval I, so that we cannot apply Parseval's equality as in the proof of Proposition 1.1. Fortunately, we have at our disposal the following celebrated generalization of Parseval's equality, due to Ingham [52]:

Theorem 4.3. *Let $(\omega_k)_{k \in K}$ be a family of* real *numbers, satisfying the uniform gap condition*

$$\gamma := \inf_{k \neq n} |\omega_k - \omega_n| > 0. \tag{4.5}$$

If I is a bounded interval I of length $|I| > 2\pi/\gamma$, then

$$\int_I |x(t)|^2 \, dt \asymp \sum_{k \in K} |x_k|^2 \tag{4.6}$$

---

[2]The case of the right endpoint is analogous.

for all functions given by the sum[3]

$$x(t) = \sum_{k \in K} x_k e^{i\omega_k t} \tag{4.7}$$

with square-summable complex coefficients x_k.

Remarks.

- Ingham's theorem improved the pioneering works of Wiener [136] and Paley–Wiener [111].
- It is natural to replace the equality by a relation $\asymp$, even in the usual case of $\omega_k = k$ for all integers. Indeed, if $I = (a, b)$ is an interval of length $|I| \geq 2\pi$, writing $2n\pi \leq |I| \leq 2(n+1)\pi$ with n an integer, we have

$$
\begin{aligned}
2k\pi \sum |x_k|^2 &= \int_a^{a+2n\pi} |x(t)|^2 \, dt \\
&\leq \int_I |x(t)|^2 \, dt \\
&\leq \int_a^{a+2(n+1)\pi} |f(t)|^2 \, dt \\
&= 2(n+1)\pi \sum |x_k|^2,
\end{aligned}
$$

and these inequalities are the best we can have for all sums (4.7).
- The above estimate contains both the so-called *direct* inequality

$$\int_I |x(t)|^2 \, dt \leq c_1 \sum_{k \in K} |x_k|^2 \tag{4.8}$$

and the so-called *inverse* inequality

$$\sum_{k \in K} |x_k|^2 \leq c_2 \int_I |x(t)|^2 \, dt. \tag{4.9}$$

The proof will show that the constants c_1 and c_2 depend only on γ and on the length of I.
- The gap condition is also *necessary* for the validity of the *inverse* inequality. Indeed, integrating the inequality

$$\left| e^{i\omega_k t} - e^{i\omega_n t} \right| \leq |(\omega_k - \omega_n)t|$$

over $I = [a, b]$ we easily deduce from (4.9) that (4.5) is satisfied with

$$\gamma := \sqrt{\frac{6}{c_2(b^3 - a^3)}}.$$

---

[3]There is no problem for the interpretation of the convergence: we can have only countably many nonzero terms, and the convergence is unconditional.

- In view of Theorem 3.1 (p. 36) it is natural to seek a more general result by considering instead of (4.7), functions of the form

$$x(t) = \sum_{k \in K} \sum_{j=0}^{m_k-1} x_{k,j} t^j e^{i\omega_k t}.$$

We shall investigate this question in Chapter 9.

In order to apply this theorem to the two series (4.3) and (4.4) above, let us note that by setting

$$\omega_{-k} := -\omega_k \quad \text{and} \quad K := \{k \in \mathbb{Z} : |k| \geq k'\}$$

for some positive integer k', the uniform gap condition (4.5) is satisfied with

$$\gamma = \gamma_{k'} := \min\Big\{2\omega_{k'}, \inf_{k \geq k'} \omega_{k+1} - \omega_k\Big\}.$$

Since

$$\omega_k = \sqrt{\Big(\frac{k\pi}{\ell}\Big)^2 + a} \to \frac{k\pi}{\ell}$$

as $k \to \infty$, we have

$$\gamma_{k'} \to \frac{\pi}{\ell},$$

and therefore the assumption $|I| > 2\ell$ implies that

$$|I| > \frac{2\pi}{\gamma_{k'}}$$

if k' is sufficiently large. Choosing such a value of k' and applying Ingham's theorem to the series (4.3) and (4.4), we conclude that

$$\int_I |u(t,0)|^2 \, dt \asymp \sum_{k=k'}^{\infty} \big(|a_k|^2 + |a_{-k}|^2\big)$$

and

$$\int_I |u_t(t,0)|^2 \, dt \asymp \sum_{k=k'}^{\infty} k^2 \big(|a_k|^2 + |a_{-k}|^2\big).$$

Hence (4.2) follows.

4.2 Proof of Ingham's Theorem

We present a simplified proof given in [7]. We first prove the theorem for *finite*[4] sums (4.7). Then the general case will be established by a standard density argument at the end of the section.

We shall use the function $H : \mathbb{R} \to \mathbb{R}$ defined by

$$H(x) := \begin{cases} \cos x & \text{if } -\pi/2 < x < \pi/2, \\ 0 & \text{otherwise,} \end{cases}$$

and its Fourier transform $h : \mathbb{R} \to \mathbb{R}$ given by

$$h(t) := \int_{-\infty}^{\infty} H(x) e^{-itx} \, dx.$$

Both functions are continuous, and H vanishes outside the open interval $(-\pi/2, \pi/2)$: see Figures 4.1 and 4.2. Let us also observe that

$$h(t) = \int_{-\pi/2}^{\pi/2} \cos x \cos tx \, dx > 0$$

for all $t \in [-1, 1]$, because $\cos x \cos tx > 0$ under the integral sign. Hence h has some positive lower bound in the interval $[-1, 1]$.[5]

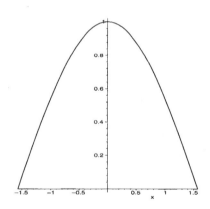
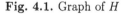

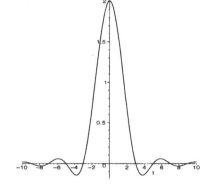

Fig. 4.1. Graph of H **Fig. 4.2.** Graph of h

Proof of the direct inequality for $\gamma = \pi$. The convolution product $G := H * H$ defines a continuous even function, vanishing outside the open interval

---

[4]Here and in the sequel a sum is called *finite* if it has at most a finite number of nonzero terms.

[5]A simple computation shows that $h(t) = \dfrac{2 \cos \pi t/2}{1 - t^2}$. Using this formula one can easily show that $h(t) \geq \pi/2$ for all $t \in [-1, 1]$.

$(-\pi, \pi)$.[6] Furthermore, its Fourier transform $g(t) = h(t)^2$ is continuous and nonnegative, and it has a positive lower bound β in the interval $[-1, 1]$.[7] See Figures 4.3 and 4.4.

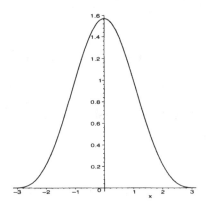

Fig. 4.3. Graph of G (direct case)   **Fig. 4.4.** Graph of g (direct case)

Now using the inverse Fourier transform we have

$$\beta \int_{-1}^{1} |x(t)|^2 \, dt \leq \int_{-\infty}^{\infty} g(t)|x(t)|^2 \, dt = 2\pi \sum_{k,n\in K} G(\omega_k - \omega_n) x_k \overline{x_n}.$$

Since $|\omega_k - \omega_n| \geq \pi$ for $k \neq n$ by the gap condition (4.5), and thus $G(\omega_k - \omega_n) = 0$, our estimate reduces to

$$\beta \int_{-1}^{1} |x(t)|^2 \, dt \leq 2\pi G(0) \sum_{k\in K} |x_k|^2.$$

This inequality remains valid for every translate $I_0 + t_0$ of $I_0 := [-1, 1]$. Indeed, putting

$$y(t) := x(t + t_0) = \sum_{k\in K} \left(x_k e^{i\omega_k t_0} \right) e^{i\omega_k t},$$

we have

$$\beta \int_{I_0+t_0} |x(t)|^2 \, dt = \beta \int_{I_0} |y(t)|^2 \, dt$$

$$\leq 2\pi G(0) \sum_{k\in K} \left| x_k e^{i\omega_k t_0} \right|^2$$

$$= 2\pi G(0) \sum_{k\in K} |x_k|^2.$$

---

[6] An easy computation shows that $2G(x) = \sin x + (\pi - x) \cos x$ for $0 \leq x \leq \pi$.
[7] We can choose $\beta = \pi^2/4$ according to the preceding footnote.

Now, every interval I can be covered by a finite number of translates $I_0+t_1,\ldots,$ $I_0 + t_m$ of I_0. Hence (4.8) follows with $c = 2m\pi G(0)/\beta$:

$$\int_I |x(t)|^2 \, dt \leq \sum_{j=1}^m \int_{I_0+t_j} |x(t)|^2 \, dt \leq \frac{2m\pi G(0)}{\beta} \sum_{k \in K} |x_k|^2.$$

Proof of the inverse inequality for $\gamma = \pi$. Choose $R > 1$ arbitrarily and set

$$G := R^2 \, H * H + H' * H'.$$

Then G is again a continuous even function, vanishing outside the interval $(-\pi, \pi)$.[8] Furthermore, its Fourier transform

$$g(t) = \left(R^2 - t^2 \right) h(t)^2$$

is continuous, negative outside the interval $I_0 = [-R, R]$, and hence bounded from above by some constant α: see Figures 4.5 and 4.6 for $R = \sqrt{2}$.

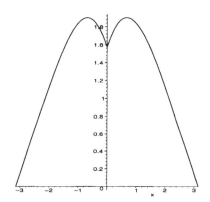
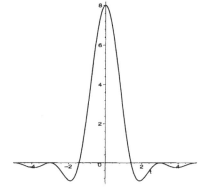

Fig. 4.5. Graph of G (inverse case)   **Fig. 4.6.** Graph of g (inverse case)

Hence

$$2\pi G(0) \sum_{k \in K} |x_k|^2 = 2\pi \sum_{k,n \in K} G(\omega_k - \omega_n) x_k \overline{x_n}$$

$$= \int_{-\infty}^{\infty} g(t)|x(t)|^2 \, dt$$

$$\leq \alpha \int_{I_0} |x(t)|^2 \, dt.$$

---

[8] An easy computation shows that $2G(x) = (R^2 + 1) \sin x + (R^2 - 1)(\pi - x) \cos x$ for $0 \leq x \leq \pi$.

Since[9] $G(0) > 0$, we conclude that

$$\sum_{k \in K} |x_k|^2 \le \frac{\alpha}{2\pi G(0)} \int_{I_0} |x(t)|^2 \, dt.$$

This estimate remains valid for every translate $I_0 + t_0$. Indeed, putting

$$y(t) := x(t + t_0) = \sum_{k \in K} \left(x_k e^{i\omega_k t_0} \right) e^{i\omega_k t}$$

again, we have

$$\sum_{k \in K} |x_k|^2 = \sum_{k \in K} \left| x_k e^{i\omega_k t_0} \right|^2$$

$$\le \frac{\alpha}{2\pi G(0)} \int_{I_0} |y(t)|^2 \, dt$$

$$= \frac{\alpha}{2\pi G(0)} \int_{I_0 + t_0} |x(t)|^2 \, dt.$$

Since every (closed) interval I of length > 2 is a translate of $[-R, R]$ for a suitable $R > 1$, the proof of (4.9) is complete.

Proof of (4.6) for finite sums in the general case. If $\gamma \ne \pi$, then, by setting

$$\omega_k' := (\pi/\gamma)\omega_k,$$

we see that the new sequence satisfies the uniform gap condition (4.5) with $\gamma' = \pi$ instead of γ, so that

$$\int_{I'} \left| \sum_{k \in K} x_k e^{i\omega_k' t'} \right|^2 \, dt' \asymp \sum_{k \in K} |x_k|^2$$

for every interval I' of length $|I'| > 2$.

Now if I is an interval of length $|I| > 2\pi/\gamma$, then the interval $I' := (\gamma/\pi)I$ satisfies $|I'| > 2$. Therefore, the desired estimates follow by a simple linear change of variable:

$$\int_I |x(t)|^2 \, dt = \int_I \left| \sum_{k \in K} x_k e^{i\omega_k t} \right|^2 \, dt = \frac{\pi}{\gamma} \int_{I'} \left| \sum_{k \in K} x_k e^{i\omega_k' t'} \right|^2 \, dt' \asymp \sum_{k \in K} |x_k|^2.$$

Proof of (4.6) for square-summable coefficients. Now consider a series of the form (4.7) with square-summable coefficients:

$$x(t) = \sum_{k \in K} x_k e^{i\omega_k t} \quad \text{with} \quad \sum_{k \in K} |x_k|^2 < \infty. \tag{4.10}$$

---

[9]Indeed, we have $G(0) = (R^2 - 1)\pi/2 > 0$ either directly from the explicit formula or because $G(0) = \int_{-\infty}^{\infty} R^2 H^2 - (H')^2 \, dx = \int_{-\pi/2}^{\pi/2} R^2 \cos^2 x - \sin^2 x \, dx.$

Let us rearrange the series so that $K = \{1, 2, \dots\}$. First we show that the series (4.7) converges in $L^2(I)$ for every interval I, so that $x(t)$ is well defined in $L^2(I)$. It suffices to show that the partial sums

$$s_p(t) = \sum_{k=1}^{p} x_k e^{i\omega_k t}$$

form a Cauchy sequence in $L^2(I)$. This follows from the already proved direct inequality, applied to the *finite* sums $s_p - s_q$, $p > q$. Indeed, we have

$$\|s_p - s_q\|^2_{L^2(I)} \le c \sum_{k=q+1}^{p} |x_k|^2,$$

and the right-hand sides tend to zero as $p, q \to \infty$ by (4.10).

Now applying (4.6) to the (finite) partial sums s_p, we have

$$\int_I \left| \sum_{k=1}^{p} x_k e^{i\omega_k t} \right|^2 dt \asymp \sum_{k=1}^{p} |x_k|^2.$$

Letting $p \to \infty$, (4.6) follows for x.

4.3 Strings with Free Endpoints II

Proposition 4.2 (p. 59) is not optimal, because we did not allow all natural initial data $u_0 \in H^1(0, \ell)$ and $u_1 \in L^2(0, \ell)$. Now we prove a more complete result:

Proposition 4.4. *Let I be an interval of length $|I| > 2\ell$. The solutions of* (4.1) *satisfy the estimates*

$$\|u(\cdot, 0)\|^2_{H^1(I)} \asymp \|u(\cdot, \ell)\|^2_{H^1(I)} \asymp E(0)$$

for all initial data $u_0 \in H^1(0, \ell)$ and $u_1 \in L^2(0, \ell)$.

Our proof will be based on a slight generalization of an important theorem due to Haraux [48]. Let $(\lambda_k)_{k \in K}$ be a family of complex numbers satisfying

$$\sup |\Re \lambda_k| < \infty,$$

and for some $k_0 \in K$ the *gap condition*

$$\gamma_0 := \inf_{k \ne k_0} |\lambda_k - \lambda_{k_0}| > 0.$$

Theorem 4.5. *Assume that for some interval I_0 we have*

$$\int_{I_0} |x(t)|^2 \, dt \asymp \sum_{k \in K \setminus \{k_0\}} |x_k|^2 \tag{4.11}$$

for all finite sums of the form

$$x(t) = \sum_{k \in K \setminus \{k_0\}} x_k e^{\lambda_k t} \tag{4.12}$$

with complex coefficients x_k. Then for every interval I of length $|I| > |I_0|$, we also have

$$\int_I |x(t)|^2 \, dt \asymp |\beta|^2 + \sum_{k \in K} |x_k|^2 \tag{4.13}$$

for all sums of the form

$$x(t) = \beta t e^{\lambda_0 t} + \sum_{k \in K} x_k e^{\lambda_k t} \tag{4.14}$$

with $\beta \in \mathbb{C}$ and with square-summable complex coefficients x_k.

Remark. Originally Haraux considered this problem with $\beta = 0$ and with purely imaginary numbers λ_k. However, his proof easily extends to the more general case considered here. A stronger generalization will be given later, in Section 6.2, p. 92.

We shall prove this theorem in the next section. Assuming its validity for the moment, let us establish a strengthening of Ingham's Theorem 4.3:

Theorem 4.6. *Let $(\omega_k)_{k \in K}$ be a family of real numbers, satisfying the uniform gap condition*

$$\gamma := \inf_{k \neq \ell} |\omega_k - \omega_\ell| > 0,$$

and set

$$\gamma' := \sup_{A \subset K} \inf_{\substack{k, \ell \in K \setminus A \\ k \neq \ell}} |\omega_k - \omega_\ell|,$$

where A runs over the finite subsets of K.

If I is a bounded interval I of length $|I| > 2\pi/\gamma'$, then

$$\int_I |x(t)|^2 \, dt \asymp \sum_{k \in K} |x_k|^2$$

for all functions given by the sum

$$x(t) = \sum_{k \in K} x_k e^{i\omega_k t}$$

with square-summable complex coefficients x_k.

Remark. Of course, we always have $\gamma' \geq \gamma$, but sometimes this inequality is strict, and this leads to an improvement of Ingham's original result by allowing *shorter* intervals I. As an illustration, let us give two simple examples:

- If $\omega_k = k$ for all integers k, then $\gamma = \gamma' = 1$.
- If $\omega_k = k^3$ for all integers k, then $\gamma = 1$ but $\gamma' = \infty$.

Proof. Given an interval $I = (a, b)$ satisfying $|I| > 2\pi/\gamma'$, choose a finite subset $A = \{k_1, \ldots, k_m\}$ of K such that $|I| > 2\pi/\gamma_A$ with

$$\gamma_A := \inf_{\substack{k,\ell \in K \setminus A \\ k \neq \ell}} |\omega_k - \omega_\ell|.$$

Furthermore, choose an interval $I_0 = (a + m\delta, b - m\delta)$ whose closure belongs to I and whose length is still strictly larger than $2\pi/\gamma_A$.

Applying Ingham's theorem, we obtain that

$$\int_{I_0} |x(t)|^2 \, dt \asymp \sum_{k \in K \setminus A} |x_k|^2$$

for all *finite* sums of the form

$$x(t) = \sum_{k \in K \setminus A} x_k e^{\omega_k t}.$$

Applying Theorem 4.5 repeatedly with $\lambda_0 = i\omega_j$ for $j = m - 1, m - 2, \ldots,$ 0, we obtain that

$$\int_{a+j\delta}^{b-j\delta} |x(t)|^2 \, dt \asymp \sum_{k \in K \setminus \{k_1, \ldots, k_j\}} |x_k|^2$$

for all *finite* sums of the form

$$x(t) = \sum_{k \in K \setminus \{k_1, \ldots, k_j\}} x_k e^{\omega_k t}.$$

For $j = 0$ this is the result we were looking for.

Now we can prove Proposition 4.4:

Proof of Proposition 4.4. For part (a) we repeat the proof of Proposition 4.2, by applying Theorem 4.6 instead of Theorem 4.3 at the end. If $\omega_k = 0$ for some k, then we conclude by applying Theorem 4.5 again.

Remark. The above proof can be easily modified to prove the following variant of Proposition 4.4: the solutions of (4.1) satisfy the estimates

$$\|u(\cdot, 0)\|_{L^2(I)}^2 \asymp \|u(\cdot, \ell)\|_{L^2(I)}^2 \asymp E_{-1}(0)$$

for all initial data $u_0 \in L^2(0, \ell)$ and $u_1 \in D^{-1}$, where we use the notation D^{-1} and $E_{-1}(t)$ introduced in Section 3.3 (p. 45). Indeed, it is equivalent to the estimate (see (4.3), p. 59)

$$\int_I \left| \sum_{k=1}^{\infty} (a_k e^{i\omega_k t} + a_{-k} e^{-i\omega_k t}) \right|^2 dt \asymp \sum_{k=1}^{\infty} k^2 (|a_k|^2 + |a_{-k}|^2),$$

established above.

More generally, infinitely many variants of Proposition 4.4 may be obtained by using the fact that if u is a solution of (4.1), then $\Delta^m u$ and the time derivative $u^{(m)}$ of u is also a solution of (4.1) for every positive integer m, with suitably modified initial data. The same remark can be made for all observability results obtained in the sequel.

4.4 Proof of Haraux's Theorem

In order to avoid the problems of convergence, first we prove the estimates (4.13) for finite sums. The general case then will follow by an easy approximation argument. By rearranging the terms if necessary, we may assume that K is the set of nonnegative integers and $k_0 = 0$.

Proof of the *direct* part of (4.13). First we note that (4.11) remains valid for every translate $I_0 + t_0$ of I_0. Indeed, putting

$$y(t) := x(t + t_0) = \sum_{k=1}^{\infty} \left(x_k e^{\lambda_k t_0} \right) e^{\lambda_k t},$$

we have

$$\int_{I_0 + t_0} |x(t)|^2 \, dt = \int_I |y(t)|^2 \, dt \asymp \sum_{k=1}^{\infty} |x_k e^{\lambda_k t_0}|^2 \asymp \sum_{k=1}^{\infty} |x_k|^2.$$

In the last step we used the boundedness of the sequence $(\Re \lambda_k)$.

Next we prove the *direct* part of (4.13), i.e., the inequality

$$\int_I |x(t)|^2 \, dt \le c \left(|\beta|^2 + \sum_{k=0}^{\infty} |x_k|^2 \right).$$

If I is a translate of I_0, then writing $I = I_0 + t_0$ and using the decomposition (4.14), we have the elementary inequality

$$\int_{I_0 + t_0} |x(t)|^2 \, dt \le 3 \int_{I_0 + t_0} \left| \beta t e^{\lambda_0 t_0} \right|^2 + \left| x_0 e^{\lambda_0 t_0} \right|^2 + \left| \sum_{k=1}^{\infty} x_k e^{\lambda_k t} \right|^2 dt$$

$$\le c(|\beta|^2 + |x_0|^2) + 3 \int_{I_0 + t_0} \left| \sum_{k=1}^{\infty} x_k e^{\lambda_k t} \right|^2 dt$$

with a suitable constant c, depending only on the interval $I_0 + t_0$. Using the assumption (4.11) for the last integral, we conclude that

$$\int_{I_0+t_0} |x(t)|^2 \, dt \le c\Big(|\beta|^2 + \sum_{k=0}^{\infty} |x_k|^2\Big)$$

with a possibly bigger constant c.

Finally, every interval I can be covered by a finite number of translates $I_1, I_2,\dots, I_n$ of I_0. Denoting the corresponding constants by $c_1,\dots,c_n$ and setting $c := c_1 + \cdots + c_n$, we have

$$\int_I |x(t)|^2 \, dt \le \sum_{j=1}^{n} \int_{I_j} |x(t)|^2 \, dt$$

$$\le \sum_{j=1}^{n} c_j \Big(|\beta|^2 + \sum_{k=0}^{\infty} |x_k|^2\Big)$$

$$= c\Big(|\beta|^2 + \sum_{k=0}^{\infty} |x_k|^2\Big).$$

Proof of the *inverse* part of (4.13). Given I with $|I| > |I_0|$, choose a translate (a, b) of I_0 and a real number $\delta > 0$ such that

$$(a - \delta, b + \delta) \subset I.$$

For x given by (4.14), the formula

$$y(t) := x(t) - \frac{1}{2\delta} \int_{-\delta}^{\delta} e^{-\lambda_0 s} x(t + s) \, ds$$

defines a function y of the form (4.12): an easy computation shows that

$$y(t) = \sum_{k=1}^{\infty} \Big[1 - \frac{\sinh(\lambda_k - \lambda_0)\delta}{(\lambda_k - \lambda_0)\delta}\Big] x_k e^{\lambda_k t} =: \sum_{k=1}^{\infty} y_k e^{\lambda_k t}.$$

As a consequence of our gap assumption $\gamma_0 > 0$ we may also assume (by slightly changing δ if necessary) that

$$\varepsilon := \inf_{k \ge 1} \Big|1 - \frac{\sinh(\lambda_k - \lambda_0)\delta}{(\lambda_k - \lambda_0)\delta}\Big|^2 > 0.$$

Then using the assumption (4.11) we have

$$\sum_{k=1}^{\infty} |x_k|^2 \le \varepsilon^{-1} \sum_{k=1}^{\infty} |y_k|^2 \le c_1 \int_a^b |y(t)|^2 \, dt \tag{4.15}$$

with a suitable constant c_1. Furthermore,

$$|y(t)|^2 \leq 2|x(t)|^2 + 2\left|\frac{1}{2\delta}\int_{-\delta}^{\delta} e^{-\lambda_0 s}x(t+s)\,ds\right|^2$$

$$\leq 2|x(t)|^2 + \frac{e^{2|\Re\lambda_0|\delta}}{\delta}\int_{-\delta}^{\delta}|x(t+s)|^2\,ds$$

$$= 2|x(t)|^2 + \frac{e^{2|\Re\lambda_0|\delta}}{\delta}\int_{t-\delta}^{t+\delta}|x(s)|^2\,ds,$$

so that

$$\int_a^b |y(t)|^2\,dt \leq 2\int_a^b |x(t)|^2\,dt + \frac{e^{2|\Re\lambda_0|\delta}}{\delta}\int_a^b\int_{t-\delta}^{t+\delta}|x(s)|^2\,ds\,dt$$

$$= 2\int_a^b |x(t)|^2\,dt + \frac{e^{2|\Re\lambda_0|\delta}}{\delta}\int_{a-\delta}^{b+\delta}\int_{\max(a,s-\delta)}^{\min(b,s+\delta)}|x(s)|^2\,dt\,ds$$

$$\leq 2\int_a^b |x(t)|^2\,dt + 2e^{2|\Re\lambda_0|\delta}\int_{a-\delta}^{b+\delta}|x(s)|^2\,ds$$

$$\leq \left(2 + 2e^{2|\Re\lambda_0|\delta}\right)\int_I |x(s)|^2\,ds.$$

Combining this result with (4.15), we conclude that

$$\sum_{k=1}^{\infty}|x_k|^2 \leq c_2\int_I |x(s)|^2\,ds \tag{4.16}$$

with a suitable constant c_2. This is still slightly weaker than the inverse part of (4.13), because β and x_0 are missing on the left-hand side. It remains to establish the estimate

$$|\beta|^2 + |x_0|^2 \leq c_3\int_I |x(s)|^2\,ds \tag{4.17}$$

with a suitable constant c_3. For this, first we note that using the decomposition (4.14), we have

$$|\beta|^2 + |x_0|^2 \leq c_4\int_I |\beta t + x_0|^2\,dt \leq 2c_4\int_I \left(|x(t)|^2 + \left|\sum_{k=1}^{\infty}x_k e^{\lambda_k t}\right|^2\right)dt$$

with some constant c_4. Since using our assumption (4.11) and then (4.16) we have

$$\int_I\left|\sum_{k=1}^{\infty}x_k e^{\lambda_k t}\right|^2\,dt \leq c_5\sum_{k=1}^{\infty}|x_k|^2 \leq 4c_2 c_5\int_I |x(s)|^2\,ds$$

with another constant c_5, we deduce from the preceding inequality that

$$|\beta|^2 + |x_0|^2 \le c_4 \int_I |\beta t + x_0|^2 \, dt \le 2c_4 \int_I |x(t)|^2 \, dt + 8c_2 c_4 c_5 \int_I |x(t)|^2.$$

Thus (4.17) holds with $c_3 := 2c_4 + 8c_2 c_4 c_5$.

End of the proof of the theorem. Now consider a series of the form (4.12) with square-summable coefficients:

$$x(t) = \beta t e^{\lambda_0 t} + \sum_{k=0}^{\infty} x_k e^{\lambda_k t}, \quad \sum_{k=0}^{\infty} |x_k|^2 < \infty. \tag{4.18}$$

First we show that the series (4.18) converges in $L^2(I)$ for every interval I, so that $x(t)$ is well-defined in $L^2(I)$. It suffices to show that the partial sums

$$s_n(t) = \beta t e^{\lambda_0 t} + \sum_{k=0}^{n} x_k e^{\lambda_k t}, \quad n = 0, 1, \ldots,$$

form a Cauchy sequence in $L^2(I)$. This follows from (4.13) applied to the *finite* sums $s_n - s_m$, $n > m$. Indeed, we have

$$\int_I |s_n(t) - s_m(t)|^2 \, dt \le c \sum_{k=m+1}^{n} |x_k|^2,$$

and the right-hand side tends to zero as $m, n \to \infty$ by (4.18).

Now applying (4.13) for the (finite) partial sums s_n, we have

$$\int_I |s_n(t)|^2 \, dt \asymp |\beta|^2 + \sum_{k=0}^{n} |x_k|^2.$$

Letting $n \to \infty$, we see that (4.13) follows for the function $x(t)$.

4.5 Strings with Fixed Endpoints or with Mixed Boundary Conditions

The proof of Proposition 4.4 can easily be adapted to other boundary conditions. Let us formulate here two variants.

4.5.1 String with Fixed Endpoints

We consider the following system, a particular case of that studied in Section 3.2 (p. 38):

$$\begin{cases} u_{tt} - u_{xx} + au = 0 & \text{in} \quad \mathbb{R} \times (0, \ell), \\ u(t, 0) = u(t, \ell) = 0 & \text{for} \quad t \in \mathbb{R}, \\ u(0, x) = u_0(x) & \text{for} \quad x \in (0, \ell), \\ u_t(0, x) = u_1(x) & \text{for} \quad x \in (0, \ell). \end{cases} \tag{4.19}$$

Setting

$$\mu_k := k\pi/\ell, \quad \omega_k := \sqrt{\mu_k^2 + a},$$

and

$$E(t) := \frac{1}{2}\int_0^\ell |u_x(t,x)|^2 + |u_t(t,x)|^2 \, dx,$$

we have the following special case of Proposition 3.2 (p. 40):

Proposition 4.7. *If*

$$u_0 \in H_0^1(0,\ell) \quad and \quad u_1 \in L^2(0,\ell),$$

then (4.19) has a unique solution satisfying

$$u \in C(\mathbb{R}; H_0^1(0,\ell)) \cap C^1(\mathbb{R}; L^2(0,\ell)).$$

It is given by a series[10]

$$u(t,x) = \sum_{k=1}^\infty (a_k e^{i\omega_k t} + a_{-k}e^{-i\omega_k t})\sin\mu_k x$$

with suitable complex coefficients a_k and a_{-k} such that

$$\sum_{k=1}^\infty k^2\big(|a_k|^2 + |a_{-k}|^2\big) < \infty.$$

Moreover, we have

$$E(0) \asymp \sum_{k=1}^\infty k^2\big(|a_k|^2 + |a_{-k}|^2\big),$$

and there exist two strictly positive functions $c_3, c_4 : \mathbb{R} \to \mathbb{R}$, independent of the choice of the initial data, such that

$$c_3(t)E(0) \le E(t) \le c_4(t)E(0)$$

for all $t \in \mathbb{R}$.

Proposition 4.8. *Given an interval I of length $|I| > 2\ell$, we have*

$$\|u_x(\cdot,0)\|_{L^2(I)}^2 \asymp \|u_x(\cdot,\ell)\|_{L^2(I)}^2 \asymp E(0)$$

for all solutions of (4.19), corresponding to arbitrary initial data

$$u_0 \in H_0^1(0,\ell) \quad and \quad u_1 \in L^2(0,\ell).$$

---

[10]If $\omega_k = 0$ for some k, then the term $a_k e^{i\omega_k t} + a_{-k}e^{-i\omega_k t}$ is replaced by $a_k + a_{-k}t$. Note that this cannot happen for more than one index k.

Proof. This is an easy adaptation of the proof of Proposition 4.4 (p. 68), by using the orthogonal basis

$$\sin \mu_k x, \quad k = 1, 2, \ldots,$$

of $L^2(0, \ell)$ and $H_0^1(0, \ell)$ instead of the corresponding cosine functions as before.

Remark. If $a = 0$ in (4.19), then we may apply Parseval's equality instead of Ingham's theorem, as in the Introduction. Therefore, Proposition 4.8 also holds in this case for intervals I having critical length $|I| = 2\ell$. The same remark applies to all results in the rest of this chapter.

4.5.2 String with Mixed Boundary Conditions

Now consider the case of one free and one fixed endpoint:

$$\begin{cases} u_{tt} - u_{xx} + au = 0 & \text{in } \mathbb{R} \times (0, \ell), \\ u(t, 0) = u_x(t, \ell) = 0 & \text{for } t \in \mathbb{R}, \\ u(0, x) = u_0(x) & \text{for } x \in (0, \ell), \\ u_t(0, x) = u_1(x) & \text{for } x \in (0, \ell). \end{cases} \tag{4.20}$$

Introducing the Hilbert spaces

$$H := L^2(0, \ell), \quad V := \{v \in H^1(0, \ell) : v(0) = 0\},$$

and setting

$$\mu_k := \left(k - \frac{1}{2}\right)\frac{\pi}{\ell}, \quad \omega_k := \sqrt{\mu_k^2 + a},$$

and

$$E(t) := \frac{1}{2}\int_0^\ell |u_x(t, x)|^2 + |u_t(t, x)|^2 \, dx,$$

we have the following special case of Proposition 3.4 (p. 47):

Proposition 4.9. *If*

$$u_0 \in V \quad and \quad u_1 \in H,$$

then (4.20) *has a unique solution satisfying*

$$u \in C(\mathbb{R}; V) \cap C^1(\mathbb{R}; H).$$

It is given by a series[11]

$$u(t, x) = \sum_{k=1}^{\infty} (a_k e^{i\omega_k t} + a_{-k} e^{-i\omega_k t}) \sin \mu_k x$$

---

[11] If $\omega_k = 0$ for some k, then the term $a_k e^{i\omega_k t} + a_{-k} e^{-i\omega_k t}$ is replaced by $a_k + a_{-k}t$. This cannot happen for more than one index k.

with suitable complex coefficients a_k and a_{-k} such that

$$\sum_{k=1}^{\infty} k^2 \left(|a_k|^2 + |a_{-k}|^2\right) < \infty.$$

Moreover, we have

$$E(0) \asymp \sum_{k=1}^{\infty} k^2 \left(|a_k|^2 + |a_{-k}|^2\right),$$

and there exist two strictly positive functions $c_3, c_4 : \mathbb{R} \to \mathbb{R}$, independent of the choice of the initial data, such that

$$c_3(t)E(0) \leq E(t) \leq c_4(t)E(0)$$

for all $t \in \mathbb{R}$.

Proposition 4.10. *Given an interval I of length $|I| > 2\ell$, we have*

$$\|u_x(\cdot, 0)\|_{L^2(I)}^2 \asymp \|u(\cdot, \ell)\|_{H^1(I)}^2 \asymp E(0)$$

for all solutions of (4.20), corresponding to arbitrary initial data $u_0 \in V$ and $u_1 \in H$.

Proof. This is another easy adaptation of the proof of Proposition 4.4 (p. 68), by using this time the orthogonal basis

$$\sin \mu_k x, \quad k = 1, 2, \ldots,$$

of H and V.

4.6 Observation at Both Ends: Free or Fixed Endpoints

It is natural to expect that a shorter observation time is sufficient if we can observe simultaneously *both* endpoints of the string. Indeed, half the time is sufficient in this case, but the proofs present unexpected new difficulties. Let us consider here the cases of two free or two fixed endpoints; the case of mixed boundary conditions will be investigated in the next section.

4.6.1 Free Endpoints

Consider again the system of Section 4.1 (p. 57):

$$\begin{cases} u_{tt} - u_{xx} + au = 0 & \text{in } \mathbb{R} \times (0, \ell), \\ u_x(t, 0) = u_x(t, \ell) = 0 & \text{for } t \in \mathbb{R}, \\ u(0, x) = u_0(x) & \text{for } x \in (0, \ell), \\ u_t(0, x) = u_1(x) & \text{for } x \in (0, \ell), \end{cases} \qquad (4.21)$$

with the energy defined by

$$E(t) := \frac{1}{2}\int_0^\ell |u(t,x)|^2 + |u_x(t,x)|^2 + |u_t(t,x)|^2 \, dx.$$

We have the following variant of Proposition 4.4 (p. 66):

Proposition 4.11. *If I is an interval of length $|I| > \ell$, then the solutions of (4.21) satisfy the estimates*

$$\|u(\cdot,0)\|_{H^1(I)}^2 + \|u(\cdot,\ell)\|_{H^1(I)}^2 \asymp E(0)$$

for all $u_0 \in H^1(0,\ell)$ and $u_1 \in L^2(0,\ell)$.

Proof. We recall from Proposition 4.1 (p. 58) that this problem has a unique solution for all initial data $u_0 \in H^1(0,\ell)$ and $u_1 \in L^2(0,\ell)$, and that it is given by the series[12]

$$u(t,x) = \sum_{k=1}^\infty (a_k e^{i\omega_k t} + a_{-k} e^{-i\omega_k t}) \cos \mu_k x,$$

with

$$\mu_k = (k-1)\pi/\ell \quad \text{and} \quad \omega_k = \sqrt{\mu_k^2 + a}$$

and with suitable complex coefficients a_k and a_{-k} such that

$$E(0) \asymp \sum_{k=1}^\infty k^2 \left(|a_k|^2 + |a_{-k}|^2\right) < \infty.$$

Thus we have to establish the estimates

$$\|u(\cdot,0)\|_{H^1(I)}^2 + \|u(\cdot,\ell)\|_{H^1(I)}^2 \asymp \sum_{k=1}^\infty k^2 \left(|a_k|^2 + |a_{-k}|^2\right). \qquad (4.22)$$

Assume first that $\omega_k \neq 0$ for all k. Putting

$$f(t) := \sum_{k=2,4,\dots} a_k e^{i\omega_k t} + a_{-k} e^{-i\omega_k t}$$

and

$$g(t) := \sum_{k=1,3,\dots} a_k e^{i\omega_k t} + a_{-k} e^{-i\omega_k t}$$

for brevity, we have the following algebraic equalities:

$$|u(t,0)|^2 + |u(0,\ell)|^2 = |f(t) + g(t)|^2 + |f(t) - g(t)|^2 = 2|f(t)|^2 + 2|g(t)|^2.$$

---

[12] If $\omega_k = 0$ for some k, then the term $a_k e^{i\omega_k t} + a_{-k} e^{-i\omega_k t}$ is replaced by $a_k + a_{-k}t$. This can happen for at most one index k.

Now applying Theorem 4.6 (p. 67), we obtain that

$$\int_I |f(t)|^2 \, dt \asymp \sum_{k=2,4,\dots} |a_k|^2 + |a_{-k}|^2$$

and

$$\int_I |g(t)|^2 \, dt \asymp \sum_{k=1,3,\dots} |a_k|^2 + |a_{-k}|^2,$$

because we have $\gamma' = 2\pi/\ell$ for both families

$$\{\pm\omega_k \ : \ k = 2, 4, \dots\} \quad \text{and} \quad \{\pm\omega_k \ : \ k = 1, 3, \dots\},$$

and $|I| > \ell = 2\pi/\gamma'$. Taking into account the above equality, we conclude that

$$\int_I |u(t,0)|^2 + |u(0,\ell)|^2 \, dt \asymp \sum_{k=1}^{\infty} |a_k|^2 + |a_{-k}|^2.$$

Using the formula

$$u_t(t,x) = \sum_{k=1}^{\infty} (i\omega_k a_k e^{i\omega_k t} - i\omega_k a_{-k} e^{-i\omega_k t}) \cos \mu_k x,$$

we obtain in a similar way that

$$\int_I |u_t(t,0)|^2 + |u_t(0,\ell)|^2 \, dt \asymp \sum_{k=1}^{\infty} |\omega_k|^2 (|a_k|^2 + |a_{-k}|^2).$$

Since $|\omega_k|^2 \asymp k^2$, by adding the last two relations we obtain (4.22).
If $\omega_k = 0$ for some k, then we conclude (4.22) again by a further application of Theorem 4.5 (p. 67).

4.6.2 Fixed Endpoints

The case of the system

$$\begin{cases} u_{tt} - u_{xx} + au = 0 & \text{in} \quad \mathbb{R} \times (0, \ell), \\ u(t,0) = u(t, \ell) = 0 & \text{for} \quad t \in \mathbb{R}, \\ u(0, x) = u_0(x) & \text{for} \quad x \in (0, \ell), \\ u_t(0, x) = u_1(x) & \text{for} \quad x \in (0, \ell), \end{cases} \tag{4.23}$$

is analogous. Setting

$$E(t) := \frac{1}{2} \int_0^\ell |u_x(t,x)|^2 + |u_t(t,x)|^2 \, dx,$$

we have the following variant of Proposition 4.8 (p. 73):

Proposition 4.12. *If I is an interval of length $|I| > \ell$, then all solutions of (4.23) satisfy the estimates*

$$\|u_x(\cdot, 0)\|_{L^2(I)}^2 + \|u_x(\cdot, \ell)\|_{L^2(I)}^2 \asymp E(0).$$

Proof. We recall from Proposition 4.7 (p. 73) that the problem (4.23) has a unique solution for all initial data $u_0 \in H_0^1(0, \ell)$ and $u_1 \in L^2(0, \ell)$, given by the series[13]

$$u(t, x) = \sum_{k=1}^{\infty} (a_k e^{i\omega_k t} + a_{-k} e^{-i\omega_k t}) \sin \mu_k x,$$

with

$$\mu_k = k\pi/\ell \quad \text{and} \quad \omega_k = \sqrt{\mu_k^2 + a},$$

and that

$$E(0) \asymp \sum_{k=1}^{\infty} k^2 (|a_k|^2 + |a_{-k}|^2).$$

Therefore, we have to establish the estimates

$$\|u_x(\cdot, 0)\|_{L^2(I)}^2 + \|u_x(\cdot, \ell)\|_{L^2(I)}^2 \asymp \sum_{k=1}^{\infty} k^2 (|a_k|^2 + |a_{-k}|^2).$$

Assume first that $\omega_k \neq 0$ for all k. Putting

$$f(t) := \sum_{k=2,4,\dots} \mu_k (a_k e^{i\omega_k t} + a_{-k} e^{-i\omega_k t})$$

and

$$g(t) := \sum_{k=1,3,\dots} \mu_k (a_k e^{i\omega_k t} + a_{-k} e^{-i\omega_k t})$$

for brevity, we have the following algebraic equalities:

$$\begin{aligned}
|u_x(t, 0)|^2 &+ |u_x(0, \ell)|^2 \\
&= \left| \sum_{k=1}^{\infty} \mu_k (a_k e^{i\omega_k t} + a_{-k} e^{-i\omega_k t}) \right|^2 \\
&\quad + \left| \sum_{k=1}^{\infty} (-1)^k \mu_k (a_k e^{i\omega_k t} + a_{-k} e^{-i\omega_k t}) \right|^2 \\
&= |f(t) + g(t)|^2 + |f(t) - g(t)|^2 \\
&= 2|f(t)|^2 + 2|g(t)|^2.
\end{aligned}$$

Hence we have to prove the relation

---

[13]If $\omega_k = 0$ for some k, then the term $a_k e^{i\omega_k t} + a_{-k} e^{-i\omega_k t}$ is replaced by $a_k + a_{-k} t$. This can happen for at most one index k.

$$\int_I |f(t)|^2 + |g(t)|^2 \, dt \asymp \sum_{k=1}^{\infty} k^2 \big(|a_k|^2 + |a_{-k}|^2\big).$$

We prove again a little more, by establishing separately that

$$\int_I |f(t)|^2 \, dt \asymp \sum_{k=2,4,\ldots} k^2 \big(|a_k|^2 + |a_{-k}|^2\big)$$

and

$$\int_I |g(t)|^2 \, dt \asymp \sum_{k=1,3,\ldots} k^2 \big(|a_k|^2 + |a_{-k}|^2\big).$$

Both relations follow by applying Theorem 4.6 (p. 67), because we have $\gamma' = 2\pi/\ell$ for both families

$$\{\pm\omega_k \ : \ k = 2, 4, \ldots\} \quad \text{and} \quad \{\pm\omega_k \ : \ k = 1, 3, \ldots\},$$

and $|I| > \ell = 2\pi/\gamma'$.

If $\omega_k = 0$ for some k, then we conclude by another application of Theorem 4.5 (p. 67).

4.7 Observation at Both Endpoints: Mixed Boundary Conditions

The case of the system

$$(4.24) \qquad \begin{cases} u_{tt} - u_{xx} + au = 0 & \text{in} \quad \mathbb{R} \times (0, \ell), \\ u(t, 0) = u_x(t, \ell) = 0 & \text{for} \quad t \in \mathbb{R}, \\ u(0, x) = u_0(x) & \text{for} \quad x \in (0, \ell), \\ u_t(0, x) = u_1(x) & \text{for} \quad x \in (0, \ell), \end{cases}$$

presents unexpected new difficulties. Setting

$$H := L^2(0, \ell) \quad \text{and} \quad V := \{v \in H^1(0, \ell) \ : \ v(0) = 0\}$$

and

$$E(t) := \frac{1}{2} \int_0^\ell |u_x(t, x)|^2 + |u_t(t, x)|^2 \, dx,$$

we have the following result:

Proposition 4.13. *If $T > \ell$, then the solutions of (4.24) satisfy the estimates*

$$\|u_x(\cdot, 0)\|^2_{L^2(0,T)} + \|u(\cdot, \ell)\|^2_{H^1(0,T)} \asymp E(0)$$

for all $u_0 \in V$ and $u_1 \in H$.

Beginning of the proof. We recall from Proposition 4.9 (p. 74) that this problem has a unique solution for all initial data $u_0 \in V$ and $u_1 \in H$. Furthermore, it is given by the series[14]

$$u(t, x) = \sum_{k=1}^{\infty} (a_k e^{i\omega_k t} + a_{-k} e^{-i\omega_k t}) \sin \mu_k x$$

with

$$\mu_k = \left(k - \frac{1}{2}\right) \frac{\pi}{\ell} \quad \text{and} \quad \omega_k = \sqrt{\mu_k^2 + a}$$

and with suitable complex coefficients a_k and a_{-k} such that

$$E(0) \asymp \sum_{k=1}^{\infty} k^2 (|a_k|^2 + |a_{-k}|^2) < \infty.$$

Thus we have to establish the estimates

$$\int_I |u_x(\cdot, 0)|^2 + |u(\cdot, \ell)|^2 + |u_t(\cdot, \ell)|^2 \ dt \asymp \sum_{k=1}^{\infty} k^2 (|a_k|^2 + |a_{-k}|^2).$$

Repeating the usual computations, we have

$$u_x(t, 0) = \sum_{k=1}^{\infty} \mu_k (a_k e^{i\omega_k t} + a_{-k} e^{-i\omega_k t}),$$

$$u(t, 0) = \sum_{k=1}^{\infty} (-1)^{k-1} (a_k e^{i\omega_k t} + a_{-k} e^{-i\omega_k t}),$$

and

$$u_t(t, 0) = \sum_{k=1}^{\infty} (-1)^{k-1} i\omega_k (a_k e^{i\omega_k t} - a_{-k} e^{-i\omega_k t}).$$

Now there are two obstacles for the application of the algebraic manipulation to separate the odd and even indices:

- the sign change between $a_k e^{i\omega_k t} - a_{-k} e^{-i\omega_k t}$ and $a_k e^{i\omega_k t} + a_{-k} e^{-i\omega_k t}$;
- the difference of the factors ω_k and μ_k.

Instead of overcoming these difficulties here by "brute force," we solve only the special case in which equation (4.24) contains no lower-order term, by applying another method. The general case will be addressed later, in Section 6.3: see Proposition 6.6 on page 103.

---

[14]If $\omega_k = 0$ for some k, then the term $a_k e^{i\omega_k t} + a_{-k} e^{-i\omega_k t}$ is replaced by $a_k + a_{-k} t$. There is at most one such k.

Proposition 4.14. *Let $a = 0$. If $T \geq \ell$, then the solutions of (4.24) satisfy the estimates*

$$\|u_x(\cdot, 0)\|^2_{L^2(0,T)} + \|u(\cdot, \ell)\|^2_{H^1(0,T)} \asymp E(0)$$

for all $u_0 \in V$ and $u_1 \in H$.

Remark. Note that the estimates also hold in the limiting case $T = \ell$.

Proof. In this case, the solution of (4.24) is given by d'Alembert's formula[15]

$$u(t, x) = f(x + t) + g(x - t)$$

with two suitable functions f and g of one real variable, depending on the initial and boundary conditions. Since the sum $f(x + t) + g(x - t)$ does not change if we replace f and g by $f + C$ and $g - C$ for some constant C, we may assume without loss of generality that

$$f(\ell) = g(\ell).$$

The boundary and initial conditions are equivalent to the following conditions on f and g:

$$\begin{cases} f(t) + g(-t) = 0 & \text{for } t \in \mathbb{R}, \\ f'(\ell + t) + g'(\ell - t) = 0 & \text{for } t \in \mathbb{R}, \\ f(x) + g(x) = u_0(x) & \text{for } x \in (0, \ell), \\ f'(x) - g'(x) = u_1(x) & \text{for } x \in (0, \ell). \end{cases}$$

In view of the equality $f(\ell) = g(\ell)$, the second condition is equivalent to

$$f(\ell + t) - g(\ell - t) = 0 \quad \text{for } t \in \mathbb{R}.$$

We use the first equality to eliminate g. Thus we obtain

$$u(t, x) = f(t + x) - f(t - x),$$

where f satisfies the following conditions:

$$\begin{cases} f(t + \ell) + f(t - \ell) = 0 & \text{for } t \in \mathbb{R}, \\ f(x) - f(-x) = u_0(x) & \text{for } x \in (0, \ell), \\ f'(x) - f'(-x) = u_1(x) & \text{for } x \in (0, \ell). \end{cases}$$

Introducing an arbitrary primitive U_1 of u_1, integrating the last condition, and combining the result with the second condition, we obtain that

---

[15] The method of d'Alembert applies only to the one-dimensional wave equation without lower-order terms. On the other hand, in this very particular case it is the simplest method in general. It can often be used to discover more general conjectures, which then could be proved by another method.

$$\begin{cases} 2f(s) = u_0(s) + U_1(s) & \text{for } s \in (0, \ell), \\ 2f(s) = -u_0(-s) + U_1(-s) & \text{for } s \in (-\ell, 0). \end{cases}$$

Furthermore, using also the first condition above, we deduce the additional relation

$$2f(s) = u_0(2\ell - s) - U_1(2\ell - s) \quad \text{for } s \in (\ell, 2\ell).$$

It follows that in the triangle[16]

$$L := \{(t, x) : x > 0 \quad \text{and} \quad x < t < \ell - x\}$$

the solution is given by the formula

$$2u(t, x) = u_0(x + t) + U_1(x + t) - u_0(t - x) - U_1(t - x),$$

while in the triangle

$$R := \{(t, x) : x < \ell \quad \text{and} \quad \ell - x < t < x\}$$

we have

$$2u(t, x) = u_0(2\ell - x - t) - U_1(2\ell - x - t) + u_0(x - t) - U_1(x - t).$$

It follows from these formulae that

$$u_x(t, 0) = u_0'(t) + u_1(t)$$

and

$$u_t(t, \ell) = -u_0'(\ell - t) + u_1(\ell - t)$$

for all $0 < t < \ell$. Hence

$$\int_0^\ell |u_x(t, 0)|^2 + |u_t(t, \ell)|^2 \, dt = \int_0^\ell |u_x(t, 0)|^2 + |u_t(\ell, \ell - t)|^2 \, dt$$

$$= \int_0^\ell |u_0' + u_1|^2 + |u_0' - u_1|^2 \, dt$$

$$= 2 \int_0^\ell |u_0'|^2 + |u_1|^2 \, dt.$$

---

[16]The letters L and R stand for "left" and "right"; make a figure.

5

Observability of Beams

In this chapter we apply the methods developed in the preceding chapter to the study of beams. One of the main differences between strings and beams is that the propagation speed in strings is *finite*, while it is *infinite* in beams. It is reflected in the results in which we determine the critical optimality time.

Throughout this chapter it will be convenient to use the notation $\mu_k := \sqrt[4]{\gamma_k}$; then $1 + \mu_k \asymp k$.

5.1 Guided Beams

First we improve and generalize Proposition 1.2 (p. 5). Fix a positive number ℓ, a real number a, and consider the one-dimensional case of the system of Section 3.6 (p. 50):

$$\begin{cases} u_{tt} + u_{xxxx} + au = 0 & \text{in} \quad \mathbb{R} \times (0, \ell), \\ u_x(t, 0) = u_{xxx}(t, 0) = u_x(t, \ell) = u_{xxx}(t, \ell) = 0 & \text{for} \quad t \in \mathbb{R}, \\ u(0, x) = u_0(x) & \text{for} \quad x \in (0, \ell), \\ u_t(0, x) = u_1(x) & \text{for} \quad x \in (0, \ell). \end{cases} \tag{5.1}$$

Introducing the Hilbert spaces $H := L^2(0, \ell)$ and

$$V := \{u \in H^2(0, \ell) \ : \ u_x(0) = u_x(\ell) = 0\}, \quad \|u\|_V := \left(\int_0^\ell |u|^2 + |u_{xx}|^2 \ dx\right)^{1/2},$$

putting

$$\mu_k := (k-1)\pi/\ell, \quad \omega_k := \sqrt{\mu_k^4 + a},$$

and

$$E(t) := \frac{1}{2} \int_0^\ell |u(t, x)|^2 + |u_{xx}(t, x)|^2 + |u_t(t, x)|^2 \ dx,$$

we have the following special case of Proposition 3.6 (p. 51):

Proposition 5.1. *If $u_0 \in V$ and $u_1 \in H$, then (5.1) has a unique solution*

$$u \in C(\mathbb{R}; V) \cap C^1(\mathbb{R}; H).$$

It is given by a series[1]

$$u(t, x) = \sum_{k=1}^{\infty} (a_k e^{i\omega_k t} + a_{-k} e^{-i\omega_k t}) \cos \mu_k x$$

with suitable complex coefficients a_k and a_{-k} such that

$$\sum_{k=1}^{\infty} k^4 (|a_k|^2 + |a_{-k}|^2) < \infty.$$

Moreover, we have

$$E(0) \asymp \sum_{k=1}^{\infty} k^4 (|a_k|^2 + |a_{-k}|^2),$$

and there exist two strictly positive functions $c_1, c_2 : \mathbb{R} \to \mathbb{R}$, independent of the choice of the initial data, such that

$$c_1(t) E(0) \le E(t) \le c_2(t) E(0)$$

for all $t \in \mathbb{R}$.

It turns out that by observing only one endpoint during an arbitrarily short time, one can already distinguish all different initial data:

Proposition 5.2. *For every fixed (arbitrarily short) interval I, all solutions of (5.1) satisfy the estimates*

$$\|u(\cdot, 0)\|_{H^1(I)}^2 \asymp \|u(\cdot, \ell)\|_{H^1(I)}^2 \asymp E(0).$$

Proof. By symmetry we consider only the case of the left endpoint. In view of the preceding proposition we have only to establish the following estimate:

$$\int_I |u(t,0)|^2 + |u_t(t,0)|^2 \, dt \asymp \sum_{k=1}^{\infty} k^4 (|a_k|^2 + |a_{-k}|^2). \qquad (5.2)$$

If none of the numbers ω_k vanishes, then $|\omega_k| \asymp k^2$, so that we may apply Theorem 4.6 (p. 67) with $\gamma' = \infty$. It follows that

$$\int_I |u(t,0)|^2 \, dt = \int_I \left| \sum_{k=1}^{\infty} a_k e^{i\omega_k t} + a_{-k} e^{-i\omega_k t} \right|^2 dt \asymp \sum_{k=1}^{\infty} |a_k|^2 + |a_{-k}|^2$$

---

[1] If $\omega_k = 0$ for some k, then the term $a_k e^{i\omega_k t} + a_{-k} e^{-i\omega_k t}$ is replaced by $a_k + a_{-k} t$.

and

$$\int_I |u_t(t,0)|^2 \, dt = \int_I \left| \sum_{k=1}^{\infty} i\omega_k a_k e^{i\omega_k t} - i\omega_k a_{-k} e^{-i\omega_k t} \right|^2 \, dt$$

$$\asymp \sum_{k=1}^{\infty} \omega_k{}^4 \big(|a_k|^2 + |a_{-k}|^2 \big).$$

Since $1 + \omega_k{}^4 \asymp k^4$, by adding the two inequalities we obtain (5.2).

If $\omega_k = 0$ for some k, then a further application of Theorem 4.5 (p. 67) yields (5.2) again.

Remark. As we explained in a remark at the end of Section 4.4 (p. 68), the above proposition admits infinitely many variants. For example, using the notation of Section 3.6 (p. 50), the solutions of (5.1) satisfy the estimates

$$\|u(\cdot,0)\|^2_{L^2(I)} \asymp \|u(\cdot,\ell)\|^2_{L^2(I)} \asymp E_{-2}(0)$$

on every interval I, for all initial data $u_0 \in D^0 = L^2(0,\ell)$ and $u_0 \in D^{-2}$.

5.2 Hinged Beams

Now we study the more realistic model of simply supported beams. Fix a positive number ℓ, a real number a, and consider the one-dimensional case of the system considered in Section 3.5 (p. 48):

$$\begin{cases} u_{tt} + u_{xxxx} + au = 0 & \text{in} \quad \mathbb{R} \times (0,\ell), \\ u(t,0) = u_{xx}(t,0) = u(t,\ell) = u_{xx}(t,\ell) = 0 & \text{for} \quad t \in \mathbb{R}, \\ u(0,x) = u_0(x) & \text{for} \quad x \in (0,\ell), \\ u_t(0,x) = u_1(x) & \text{for} \quad x \in (0,\ell). \end{cases} \tag{5.3}$$

Putting

$$\mu_k := k\pi/\ell, \quad \omega_k := \sqrt{\mu_k^4 + a},$$

and

$$E_{-1}(t) := \frac{1}{2} \int_0^\ell |u_x(t,x)|^2 + |(\Delta^{-1} u_t(t,x))_x|^2 \, dx,$$

we have the following special case of the variant $s = -1$ of Proposition 3.5 (p. 48; see also the end of Section 3.5, p. 50):

Proposition 5.3. *If* $u_0 \in H_0^1(0,\ell)$ *and* $u_1 \in H^{-1}(0,\ell)$, *then* (5.3) *has a unique solution*

$$u \in C(\mathbb{R}; H_0^1(0,\ell)) \cap C^1(\mathbb{R}; H^{-1}(0,\ell)).$$

It is given by a series[2]

$$u(t,x) = \sum_{k=1}^{\infty} (a_k e^{i\omega_k t} + a_{-k} e^{-i\omega_k t}) \sin \mu_k x$$

with suitable complex coefficients a_k and a_{-k} such that

$$\sum_{k=1}^{\infty} k^2 (|a_k|^2 + |a_{-k}|^2) < \infty.$$

Moreover, we have

$$E_{-1}(0) \asymp \sum_{k=1}^{\infty} k^2 (|a_k|^2 + |a_{-k}|^2).$$

and there exist two strictly positive functions $c_1, c_2 : \mathbb{R} \to \mathbb{R}$, independent of the choice of the initial data, such that

$$c_1(t) E_{-1}(0) \le E_{-1}(t) \le c_2(t) E_{-1}(0)$$

for all $t \in \mathbb{R}$.

We have again observability by using only one endpoint and for an arbitrarily short time interval:

Proposition 5.4. *For every fixed (arbitrarily short) interval I, all solutions of (5.3) satisfy the estimates*

$$\|u_x(\cdot,0)\|^2_{L^2(I)} \asymp \|u_x(\cdot,\ell)\|^2_{L^2(I)} \asymp E_{-1}(0).$$

Proof. As above, by symmetry we consider only the case of the left endpoint. In view of the preceding proposition we have to establish the following estimate:

$$\int_I |u_x(t,0)|^2 \, dt \asymp \sum_{k=1}^{\infty} k^2 (|a_k|^2 + |a_{-k}|^2). \tag{5.4}$$

If $\omega_k \ne 0$ for all k, then we may apply Theorem 4.6 (p. 67) with $\gamma' = \infty$ because $|\omega_k| \asymp k^2$; it follows that

$$\int_I |u_x(t,0)|^2 \, dt = \int_I \left| \sum_{k=1}^{\infty} \mu_k (a_k e^{i\omega_k t} + a_{-k} e^{-i\omega_k t}) \right|^2 \, dt \asymp \sum_{k=1}^{\infty} k^2 (|a_k|^2 + |a_{-k}|^2).$$

If $\omega_k = 0$ for some k, then (5.4) follows again by a further application of Theorem 4.5 (p. 67).

Remark. There again exist many variants of this proposition. For example, with the notation of Section 3.5 (p. 48), the solutions of (5.3) satisfy the estimates

$$\|u_{xxx}(\cdot,0)\|^2_{L^2(I)} \asymp \|u_{xxx}(\cdot,\ell)\|^2_{L^2(I)} \asymp E_1(0)$$

on every interval I, for all initial data $u_0 \in D^3$ and $u_0 \in D^1$.

---

[2] If $\omega_k = 0$ for some k, then the term $a_k e^{i\omega_k t} + a_{-k} e^{-i\omega_k t}$ is replaced by $a_k + a_{-k} t$.

5.3 Mixed Boundary Conditions

Consider finally the case of mixed boundary conditions, a special case of the system studied in Section 3.7 (p. 52):

$$\begin{cases} u_{tt} + u_{xxxx} + au = 0 & \text{in} \quad \mathbb{R} \times (0, \ell), \\ u(t, 0) = u_{xx}(t, 0) = u_x(t, \ell) = u_{xxx}(t, \ell) = 0 & \text{for} \quad t \in \mathbb{R}, \\ u(0, x) = u_0(x) & \text{for} \quad x \in (0, \ell), \\ u_t(0, x) = u_1(x) & \text{for} \quad x \in (0, \ell). \end{cases} \tag{5.5}$$

Putting

$$\mu_k := \left(k - \frac{1}{2}\right)\frac{\pi}{\ell},$$

$$\omega_k := \sqrt{\mu_k^4 + a},$$

$$E(t) := \frac{1}{2} \int_0^\ell |u(t, x)|^2 + |u_{xx}(t, x)|^2 + |u_t(t, x)|^2 \, dx,$$

and introducing the Hilbert spaces $H = L^2(0, \ell)$ and

$$V := \{v \in H^2(0, \ell) \; : \; v(0) = v'(\ell) = 0\}, \quad \|v\|_V := \left(\int_0^\ell |v|^2 + |v_{xx}|^2 \, dx\right)^{1/2},$$

we have the following special case of Proposition 3.7 (p. 52):

Proposition 5.5. *If $u_0 \in V$ and $u_1 \in H$, then (5.5) has a unique solution*

$$u \in C(\mathbb{R}; V) \cap C^1(\mathbb{R}; H).$$

It is given by a series[3]

$$u(t, x) = \sum_{k=1}^\infty (a_k e^{i\omega_k t} + a_{-k} e^{-i\omega_k t}) \sin \mu_k x$$

with suitable complex coefficients a_k and a_{-k} such that

$$\sum_{k=1}^\infty k^4 \big(|a_k|^2 + |a_{-k}|^2\big) < \infty.$$

Moreover, we have

$$E(0) \asymp \sum_{k=1}^\infty k^4 \big(|a_k|^2 + |a_{-k}|^2\big),$$

---

[3] If $\omega_k = 0$ for some k, then the term $a_k e^{i\omega_k t} + a_{-k} e^{-i\omega_k t}$ is replaced by $a_k + a_{-k}t$.

and there exist two strictly positive functions $c_1, c_2 : \mathbb{R} \to \mathbb{R}$, independent of the choice of the initial data, such that

$$c_1(t)E(0) \le E(t) \le c_2(t)E(0)$$

for all $t \in \mathbb{R}$.

We have analogous observability results as in the preceding cases:

Proposition 5.6. *For every fixed (arbitrarily short) interval I, all solutions of (5.1) satisfy the estimates*

$$\|u_x(\cdot, 0)\|_{L^2(I)}^2 \asymp \|u(\cdot, \ell)\|_{H^1(I)}^2 \asymp E(0).$$

Proof. It is a straightforward adaptation of the proofs of Propositions 5.2 and 5.4, so it is left to the reader.

6

Vector Sum Estimates

We left some problems of the preceding two chapters unsolved. In order to solve them in an elegant way, we first generalize the main abstract Theorems 4.3 and 4.5 of Ingham and Haraux to the case of vector coefficients. They will also prove to be useful in various other observability problems. The first two sections of this chapter contain these generalizations, while the remainder of the chapter is devoted to applications. In order to make our book self-contained, at the end of the chapter we give a short introduction to the celebrated multiplier method, by presenting simplified proofs of two important theorems of J.-L. Lions on the observability of membranes and plates.

6.1 An Ingham-Type Theorem for Vector-Valued Functions

Let $(\omega_k)_{k \in K}$ be a family of real numbers satisfying for some positive integer M and for some real number $\gamma > 0$ the following *weakened gap condition*:

no interval $(\omega_k - \gamma, \omega_k + \gamma)$ contains more

than M members of the family (ω_k). (6.1)

Let $(E_k)_{k \in K}$ be a family of vectors in a complex Hilbert space $\mathcal{H}$, and denote by $\mathcal{Z}$ the linear hull of these vectors. Let $p(\cdot, \cdot)$ be a given semiscalar product (positive semidefinite sesquilinear form) on $\mathcal{Z}$, and denote by $p(\cdot)$ the corresponding seminorm.

Theorem 6.1. *Assume* (6.1). *For every bounded interval I of length $|I| > 2\pi/\gamma$, there exists a number $\eta > 0$ such that if*

$$|p(E_k, E_n)| \le \eta p(E_k) p(E_n) \quad \text{whenever} \quad |\omega_k - \omega_n| < \gamma \quad \text{but} \quad k \ne n, \quad (6.2)$$

then all finite *sums*

$$U(t) = \sum_{k \in K} U_k e^{i\omega_k t} E_k, \quad U_k \in \mathbb{C},$$

satisfy the estimates

$$\int_I p(U(t))^2 \, dt \asymp \sum_{k \in K} |U_k|^2 p(E_k)^2. \tag{6.3}$$

Remarks.

- For $M = 1$ the condition (6.2) is empty, so that we do not need any η. If $X = \mathbb{C}$ and $p(x, y) = x\bar{y}$, then this result reduces to Ingham's original theorem (p. 59).
- In comparison with Ingham's theorem, for $M > 1$ the weakening of the gap condition is compensated by the *quasi-orthogonality* of the coefficients corresponding to close (or equal) exponents. The proof will provide an explicit value of η: setting $R := |I|\gamma/(2\pi)$ and denoting by β the maximum of the function $(R^2 + 1)\sin x + (R^2 - 1)(\pi - x)$ on the interval $[0, \pi]$, we may choose an η such that $(M - 1)\beta\eta < (R^2 - 1)\pi$. As we shall see later in this chapter, this quasi-orthogonality property is often satisfied in the applications. This was first observed in [76].
- We consider only finite sums, since p is assumed to be defined only on $\mathcal{Z}$. However, once the estimate (6.3) is established, a standard density argument allows us to extend the integrals in (6.3) to more general functions, so that (6.3) is valid for all sums with square-summable coefficients.
- This theorem is stronger than Theorem 1.8 in [76], because here the quasi-orthogonality is required for fewer pairs of indices. This will be achieved by applying Ingham's "second method" instead of the first one, used before.
- An even more general result is established in [9], which covers the case of generalized eigenvectors. See also [101] for an application in which infinitely many generalized eigenvectors appear in a natural way.

The proof is an adaptation of the proof of Theorem 4.3. By the same arguments as there,

- we may assume without loss of generality that $\gamma = \pi$;
- it suffices to establish the direct inequality for one, arbitrarily short, interval;
- it suffices to prove the inverse inequality for the intervals $(-R, R)$ with $R > 1$.

Let us choose the same functions H and h as in the proof of Theorem 4.3,

$$H(x) := \begin{cases} \cos x & \text{if } -\pi/2 < x < \pi/2, \\ 0 & \text{otherwise,} \end{cases}$$

and its Fourier transform $h : \mathbb{R} \to \mathbb{R}$ given by

$$h(t) := \int_{-\infty}^{\infty} e^{-itx} H(x) \, dx = \frac{2 \cos \pi t/2}{1 - t^2}.$$

In the proof we assume without loss of generality that $p(E_k) = 1$ for all k. Indeed, terms with $p(E_k) = 0$ do not contribute to either side of (6.3), while the other terms can be normalized. Then we have

$$|p(E_k, E_n)| \leq 1 \quad \text{for all} \quad k, n,$$

and (6.2) takes the form

$$|p(E_k, E_n)| \leq \eta \quad \text{whenever} \quad |\omega_k - \omega_n| < \gamma \quad \text{but} \quad k \neq n.$$

Proof of the direct equality. We recall that $G := H * H$ is a continuous function, vanishing outside $(-\pi, \pi)$ and attaining its maximum at 0, its Fourier transform g is continuous and nonnegative, and $g \geq 1$ on some small interval $(-r, r)$. Therefore, we have the following estimates:

$$\int_{-r}^{r} p(U(t))^2 \, dt \leq \int_{-\infty}^{\infty} g(t) p(U(t))^2 \, dt$$

$$= 2\pi \sum_{k,n \in K} U_k \overline{U_n} G(\omega_k - \omega_n) p(E_k, E_n)$$

$$= 2\pi \sum_{\substack{k,n \in K: \\ |\omega_k - \omega_n| < \gamma}} U_k \overline{U_n} G(\omega_k - \omega_n) p(E_k, E_n)$$

$$\leq 2\pi G(0) \sum_{\substack{k,n \in K: \\ |\omega_k - \omega_n| < \gamma}} \frac{|U_k|^2 + |U_n|^2}{2}$$

$$\leq 2\pi G(0) M \sum_{k \in K} |U_k|^2.$$

In the last step we used that by assumption (6.1), no term $|U_k|^2$ appears more than M times in the sum.

Proof of the inverse equality. Now we recall that $G := R^2 H * H + H' * H'$ is a continuous function, vanishing outside $(-\pi, \pi)$, satisfying $G(0) > 0$, and thus having a positive maximum β, while its Fourier transform g is continuous, negative outside $[-R, R]$, and hence bounded from above by some constant α. Therefore, we have the following estimates:

$$\alpha \int_{-R}^{R} p(U(t))^2 \, dt \geq \int_{-\infty}^{\infty} g(t) p(U(t))^2 \, dt$$

$$= 2\pi \sum_{k,n \in K} U_k \overline{U_n} G(\omega_k - \omega_n) p(E_k, E_n)$$

$$= 2\pi \sum_{\substack{k,n \in K: \\ |\omega_k - \omega_n| < \gamma}} U_k \overline{U_n} G(\omega_k - \omega_n) p(E_k, E_n)$$

$$= 2\pi G(0) \Big(\sum_{k \in K} |U_k|^2 \Big) + 2\pi \sum_{\substack{k,n \in K: \\ |\omega_k - \omega_n| < \gamma, \\ k \neq n}} U_k \overline{U_n} G(\omega_k - \omega_n) p(E_k, E_n)$$

$$\geq 2\pi G(0) \Big(\sum_{k \in K} |U_k|^2 \Big) - 2\pi \beta \eta \sum_{\substack{k,n \in K: \\ |\omega_k - \omega_n| < \gamma, \\ k \neq n}} |U_k \overline{U_n}|$$

$$\geq 2\pi G(0) \Big(\sum_{k = -\infty}^{\infty} |U_k|^2 \Big) - 2\pi \beta \eta \sum_{\substack{|\omega_k - \omega_n| < \gamma \\ k \neq n}} \frac{|U_k|^2 + |U_n|^2}{2}$$

$$\geq 2\pi G(0) \Big(\sum_{k \in K} |U_k|^2 \Big) - 2\pi (M - 1) \beta \eta \Big(\sum_{k \in K} |U_k|^2 \Big)$$

$$= 2\pi \big(G(0) - (M - 1) \beta \eta \big) \sum_{k \in K} |U_k|^2.$$

With a sufficiently small $\eta > 0$ such that $(M - 1)\beta\eta < G(0)$, the desired inequality follows.

6.2 An Haraux-Type Theorem for Vector-Valued Functions

We are going to generalize Theorem 4.5 of Haraux (p. 67). As in Chapter 3, we consider the solutions of the problem

$$U' = \mathcal{A}U, \quad U(0) = U_0, \tag{6.4}$$

in an infinite-dimensional *complex* Hilbert space $\mathcal{H}$, where $\mathcal{A}$ is an unbounded linear operator defined on some linear subspace of $\mathcal{H}$, with values in $\mathcal{H}$, satisfying hypothesis (RB) on page 34. We recall[1] that the solutions of (6.4) are given by the series

$$U(t) = \sum_{k \in K} \sum_{\ell=1}^{m_k} U_{k,\ell} F_{k,\ell}(t) \quad \text{with} \quad F_{k,\ell}(t) := \sum_{j=0}^{\ell-1} \frac{t^j e^{\lambda_k t}}{j!} E_{k,\ell-j},$$

where the complex numbers $U_{k,\ell}$ are the coefficients of the initial value U_0 in the Riesz basis:

---

[1]See Theorem 3.1, p. 36.

$$U_0 = \sum_{k \in K} \sum_{\ell=1}^{m_k} U_{k,\ell} E_{k,\ell}.$$

As in the preceding section, we consider only finite sums, i.e., solutions corresponding to initial data belonging to the linear hull $\mathcal{Z}$ of the Riesz basis; we recall that in this case the solutions are of class C^∞. For every *finite* subset A of K, it will be useful to denote by $\mathcal{Z}_A$ the linear hull of the basis vectors $E_{k,\ell}$ with $k \in K \setminus A$ and $\ell = 1, \ldots, m_k$; then $\mathcal{Z}_A$ is a finite-codimensional subspace of $\mathcal{Z}$. Notice that if $U_0 \in \mathcal{Z}_A$, then the solution of (6.4) also has its values in $\mathcal{Z}_A$, and we have

$$U(t) = \sum_{k \in K \setminus A} \sum_{\ell=1}^{m_k} U_{k,\ell} F_{k,\ell}(t) \quad \text{with} \quad F_{k,\ell}(t) := \sum_{j=0}^{\ell-1} \frac{t^j e^{\lambda_k t}}{j!} E_{k,\ell-j}$$

and

$$U_0 = \sum_{k \in K \setminus A} \sum_{\ell=1}^{m_k} U_{k,\ell} E_{k,\ell}.$$

Now we are ready to state our theorem, first obtained in [77]:

Theorem 6.2. *Let us be given a finite number of seminorms $p_1, \ldots, p_m$ on $\mathcal{Z}$. Assume that*

(i) There exist a finite subset A of K and intervals $I_1, \ldots, I_m$ such that the estimates

$$\sum_{j=1}^{m} \int_{I_j} p_j(U(t))^2 \, dt \asymp \|U_0\|^2 \tag{6.5}$$

hold for all solutions of (6.4) with $U_0 \in \mathcal{Z}_A$.

(ii) There exist intervals $I_1', \ldots, I_m'$ such that the estimates

$$\sum_{j=1}^{m} \int_{I_j'} p_j(U(t))^2 \, dt \asymp \|U_0\|^2 \tag{6.6}$$

hold for all solutions of (6.4) with $U_0 \in \mathcal{Z}$ such that $\mathcal{A}U_0 = \lambda_k U_0$ for some $k \in A$.

Then for any choice of intervals $J_1, \ldots, J_m$ such that each J_j contains the closure of I_j in its interior, the estimates

$$\sum_{j=1}^{m} \int_{J_j} p_j(U(t))^2 \, dt \asymp \|U_0\|^2 \tag{6.7}$$

hold for all solutions of (6.4) with $U_0 \in \mathcal{Z}$.

Remarks.

- The use of several seminorms corresponds to various different types of observations, at different times.
- Notice that the lengths of the intervals I'_j do not affect the choice of the intervals J_j in the conclusion. This is natural, because if hypothesis (ii) is satisfied for some choice of the intervals I'_j, then it is also satisfied for any other choice. Indeed, hypothesis (ii) is equivalent to the following *uniqueness property*, where the intervals I'_j no longer appear: if $\mathcal{A}U_0 = \lambda_k U_0$ for some $k \in A$ and if $p_1(U_0) = \cdots = p_m(U_0) = 0$, then $U_0 = 0$. The equivalence follows from the following relation, where $U(t) = e^{\lambda_k t}U_0$ denotes the solution of (6.4):

$$\sum_{j=1}^{m} \int_{I'_j} p_j(U(t))^2 \, dt = \sum_{j=1}^{m} \left(\int_{I'_j} \left| e^{\lambda_k t} \right|^2 \, dt \right) p_j(U_0)^2. \tag{6.8}$$

  Indeed, if hypothesis (ii) is satisfied and $p_1(U_0) = \cdots = p_m(U_0) = 0$, then the right-hand side vanishes; hence (6.6) and (6.8) imply that $U_0 = 0$. Conversely, the uniqueness property implies that the square root of the left-hand side of (6.8) is a norm on the *finite-dimensional* subspace spanned by the eigenvectors figuring in hypothesis (ii), so that it is equivalent to the norm of $\mathcal{H}$ on this subspace.
- Let us emphasize a peculiar feature of the hypotheses of the theorem: if A is replaced by a bigger set, then the hypotheses are still satisfied; moreover, since a smaller set of initial values is considered, the estimates can hold with even *shorter* intervals I_j. This crucial observation often makes it possible to improve drastically the estimates of the sufficient observability time. See, e.g., [67], where many applications are given by using a special case of the above theorem.

The proof consists of several steps.

Proof of the direct part of (6.7). We are going to establish the estimates[2]

$$\sum_{j=1}^{m} \int_{J_j} p_j(U(t))^2 \, dt \le c\|U_0\|^2 \tag{6.9}$$

for all $U_0 \in \mathcal{Z}$ and for all choices of the intervals J_j. Let us begin with the case $J_j = I_j$.

Given $U_0 \in \mathcal{Z}$ arbitrarily, we write the solution of (6.4) in the form $U = V + W$ with

$$V(t) = \sum_{k \in A} \sum_{\ell=1}^{m_k} U_{k,\ell} F_{k,\ell}(t), \quad V_0 = V(0),$$

---

[2]Here and in the sequel the letter c denotes various constants, independent of the particular choice of the initial data.

and

$$W(t) = \sum_{k \in K \backslash A} \sum_{\ell=1}^{m_k} U_{k,\ell} F_{k,\ell}(t), \quad W_0 = W(0).$$

Applying our assumption (6.5) for W instead of U, we have

$$\sum_{j=1}^{m} \int_{I_j} p_j(W(t))^2 \, dt \le c\|W_0\|^2. \tag{6.10}$$

Furthermore, since V is given by a finite sum, we also have

$$\sum_{j=1}^{m} \int_{I_j} p_j(V(t))^2 \, dt \le c\|V_0\|^2. \tag{6.11}$$

Indeed, using the definition of the seminorms we have

$$\sum_{j=1}^{m} \int_{I_j} p_j(V(t))^2 \, dt = \sum_{j=1}^{m} \int_{I_j} p_j \Big(\sum_{k \in A} \sum_{\ell=1}^{m_k} U_{k,\ell} F_{k,\ell}(t) \Big)^2 \, dt$$

$$\le \sum_{j=1}^{m} \int_{I_j} \Big(\sum_{k \in A} \sum_{\ell=1}^{m_k} |U_{k,\ell}| \sum_{j=0}^{\ell-1} \frac{|t|^j e^{\lambda_k t}}{j!} p_j(E_{k,\ell-j}) \Big)^2 \, dt$$

$$\le c \sum_{k \in A} \sum_{\ell=1}^{m_k} |U_{k,\ell}|^2$$

$$\le c\|V_0\|^2.$$

Now using the triangle inequality and then the Riesz basis property we conclude from (6.10) and (6.11) that

$$\sum_{j=1}^{m} \int_{I_j} p_j(U(t))^2 \, dt \le 2 \sum_{j=1}^{m} \int_{I_j} p_j(V(t))^2 \, dt + 2 \sum_{j=1}^{m} \int_{I_j} p_j(W(t))^2 \, dt$$

$$\le 2c\|V_0\|^2 + 2c\|W_0\|^2$$

$$\le c \sum_{k \in K} \sum_{\ell=1}^{m_k} |U_{k,\ell}|^2$$

$$\le c\|U_0\|^2.$$

Now given m intervals $J_1, \dots, J_m$ arbitrarily, let us choose finitely many real numbers $t_1, \dots, t_n$ such that for each $j = 1, \dots, m$ the translates

$$I_j + t_1, \dots, I_j + t_n$$

of the interval I_j cover the interval J_j. Then, taking into account the translation invariance of equation (6.4), we have

$$\sum_{j=1}^{m} \int_{J_j} p_j(U(t))^2 \ dt \leq \sum_{i=1}^{n} \sum_{j=1}^{m} \int_{I_j+t_i} p_j(U(t))^2 \ dt \leq c \sum_{i=1}^{n} \|U(t_i)\|^2.$$

The proof of (6.9) will be complete if we establish for every fixed $t \in \mathbb{R}$ the estimate

$$\|U(t)\|^2 \leq c(t)\|U_0\|^2.$$

This follows by a direct computation: using the Riesz basis property, we have

$$\|U(t)\|^2 = \left\| \sum_{k \in K} \sum_{i=1}^{m_k} \left(\sum_{\ell=i}^{m_k} U_{k,\ell} \frac{t^{\ell-i} e^{\lambda_k t}}{(\ell-i)!} \right) E_{k,i} \right\|^2 \leq c \sum_{k \in K} \sum_{i=1}^{m_k} \left| \sum_{\ell=i}^{m_k} U_{k,\ell} \frac{t^{\ell-i} e^{\lambda_k t}}{(\ell-i)!} \right|^2.$$

Since the sequences (m_k) and $(\Re \lambda_k)$ are bounded, the last expression is majorized by

$$c(t) \sum_{k \in K} \sum_{i=1}^{m_k} |U_{k,i}|^2,$$

and we conclude by applying the Riesz basis property again.

The inverse inequality in (6.7) is considerably deeper. We begin with a technical lemma. Given $\delta > 0$ and $\lambda \in \mathbb{C}$ arbitrarily, for every continuous function $U : \mathbb{R} \to \mathcal{H}$ we define, following Haraux [48], another continuous function $I_{\delta,\lambda}U : \mathbb{R} \to \mathcal{H}$ by the formula

$$I_{\delta,\lambda}U(t) = U(t) - \frac{1}{2\delta} \int_{-\delta}^{\delta} e^{-\lambda s} U(t+s) \ ds.$$

Lemma 6.3. (a) Given a solution

$$U(t) = \sum_{k \in K} \sum_{\ell=1}^{m_k} U_{k,\ell} F_{k,\ell}(t)$$

of (6.4), $V := I_{\delta,\lambda}U$ is also a solution of (6.4) (with a different initial datum V_0):

$$V(t) = \sum_{k \in K} \sum_{\ell=1}^{m_k} V_{k,\ell} F_{k,\ell}(t).$$

If $U_0 \in \mathcal{Z}$, then $V_0 \in \mathcal{Z}$.

(b) If p is a seminorm on $\mathcal{Z}$ and (a,b) is an interval, then we have the estimate

$$\int_a^b p(V(t))^2 \ dt \leq c \int_{a-\delta}^{b+\delta} p(U(t))^2 \ dt, \quad \text{for all} \quad U_0 \in \mathcal{Z}. \tag{6.12}$$

(c) We have $V_{k,m_k} = 0$ whenever $\lambda = \lambda_k$.

(d) For all but countably many $\delta > 0$ we have the estimate

$$\sum_{\substack{k \in K \\ \lambda \neq \lambda_k}} \sum_{\ell=1}^{m_k} |U_{k,\ell}|^2 \leq c \sum_{\substack{k \in K \\ \lambda \neq \lambda_k}} \sum_{\ell=1}^{m_k} |V_{k,\ell}|^2. \tag{6.13}$$

Remark. The property (c) is crucial: it shows that by applying the operator $I_{\delta,\lambda}$ we can remove at least one term from the sum defining $U(t)$.

Proof. (a) Using the translation invariance of equation (6.4) we have

$$V' = (I_{\delta,\lambda}U)' = I_{\delta,\lambda}U' = I_{\delta,\lambda}\mathcal{A}U = \mathcal{A}(I_{\delta,\lambda}U) = \mathcal{A}V.$$

The second relation follows from the $\mathcal{A}$-invariance of Z.

(b) For every fixed $t \in \mathbb{R}$ we have

$$p(V(t))^2 \leq 2p(U(t))^2 + 2p\left(\frac{1}{2\delta}\int_{-\delta}^{\delta} e^{-\lambda s}U(t+s)\,ds\right)^2$$

$$\leq 2p(U(t))^2 + \frac{1}{2\delta^2}\left(\int_{-\delta}^{\delta} e^{-\lambda s}p(U(t+s))\,ds\right)^2$$

$$\leq 2p(U(t))^2 + \frac{1}{2\delta^2}\int_{-\delta}^{\delta}|e^{-\lambda s}|^2\,ds\int_{-\delta}^{\delta}p(U(t+s))^2\,ds$$

$$\leq 2p(U(t))^2 + \frac{e^{2|\Re\lambda|\delta}}{\delta}\int_{t-\delta}^{t+\delta}p(U(s))^2\,ds.$$

Therefore,

$$\int_a^b p(V(t))^2\,dt \leq 2\int_a^b p(U(t))^2\,dt + \frac{e^{2|\Re\lambda|\delta}}{\delta}\int_a^b\int_{t-\delta}^{t+\delta}p(U(s))^2\,ds\,dt$$

$$= 2\int_a^b p(U(t))^2\,dt + \frac{e^{2|\Re\lambda|\delta}}{\delta}\int_{a-\delta}^{b+\delta}\int_{\max\{a,s-\delta\}}^{\min\{b,s+\delta\}}p(U(s))^2\,dt\,ds$$

$$\leq 2\int_a^b p(U(t))^2\,dt + 2e^{2|\Re\lambda|\delta}\int_{a-\delta}^{b+\delta}p(U(s))^2\,dt,$$

and (6.12) follows with

$$c = 2 + 2e^{2|\Re\lambda|\delta}.$$

(c) We have

$$V(t) = U(t) - \sum_{k \in K}\sum_{i=1}^{m_k}\sum_{\ell=i}^{m_k}\frac{U_{k,\ell}}{2\delta(\ell-i)!}\left(\int_{-\delta}^{\delta} e^{-\lambda s}(t+s)^{\ell-i}e^{\lambda_k(t+s)}\,ds\right)E_{k,i},$$

whence

$$V_0 = U_0 - \sum_{k \in K}\sum_{i=1}^{m_k}\sum_{\ell=i}^{m_k}\frac{U_{k,\ell}}{2\delta(\ell-i)!}\left(\int_{-\delta}^{\delta} s^{\ell-i}e^{(\lambda_k-\lambda)s}\,ds\right)E_{k,i}.$$

It follows that

$$V_{k,i} = U_{k,i} - \sum_{\ell=i}^{m_k} \frac{U_{k,\ell}}{2\delta(\ell-i)!} \int_{-\delta}^{\delta} e^{(\lambda_k-\lambda)s} s^{\ell-i} \, ds \qquad (6.14)$$

for all k, i. In particular,

$$V_{k,m_k} = U_{k,m_k}\left(1 - \frac{1}{2\delta}\int_{-\delta}^{\delta} e^{(\lambda_k-\lambda)s} \, ds\right),$$

and the last factor vanishes if $\lambda_k = \lambda$.

(d) It follows from (6.14) that the matrix A_k that tranforms the vector

$$(U_{k,1},\ldots,U_{k,m_k}) \quad \text{into} \quad (V_{k,1},\ldots,V_{k,m_k})$$

is triangular, with determinant equal to

$$\Delta_k := \prod_{i=1}^{m_k}\left(1 - \frac{1}{2\delta}\int_{-\delta}^{\delta} e^{((\lambda_k-\lambda)s)} \, ds\right) = \prod_{i=1}^{m_k}\left(1 - \frac{\sinh(\lambda_k-\lambda)\delta}{(\lambda_k-\lambda)\delta}\right).$$

This is an analytic function of δ for every k, and it is not identically zero if $\lambda \neq \lambda_k$. Hence for all but countably many $\delta > 0$ we have $\Delta_k \neq 0$ whenever $\lambda_k \neq \lambda$, so that A_k is invertible. Furthermore, for every such fixed δ, we deduce from the assumption $|\lambda_k| \to \infty$ and from the boundedness of the sequences (m_k) and $(\Re\lambda_k)$ that $\|I - A_k\| \to 0$. Indeed, observe that all integrals in (6.14) tend to zero as $k \to \infty$: integrate by parts or use the Riemann–Lebesgue lemma. Therefore, we also have $\|I - A_k^{-1}\| \to 0$ as $k \to \infty$. Hence

$$\sup_{\lambda_k \neq \lambda} \|I - A_k^{-1}\| < \infty,$$

and (6.13) follows.

Now we establish a somewhat weaker form of the inverse inequality in (6.7): observe that the terms $k \in A$ are missing on the left-hand side of inequality (6.15) below.

Lemma 6.4. *Under the assumptions of Theorem 6.2 we have*

$$\sum_{k \in K \setminus A} \sum_{\ell=1}^{m_k} |U_{k,\ell}|^2 \leq c \sum_{j=1}^{m} \int_{J_j} p_j(U(t))^2 \, dt \qquad (6.15)$$

for all solutions of (6.4) with $U_0 \in \mathcal{Z}$.

Proof. Set

$$M = \sum_{k \in A} m_k$$

and fix a sufficiently small $\delta > 0$ so that writing $I_j = (a_j, b_j)$, we have

$$(a_j - M\delta, b_j + M\delta) \subset J_j \quad \text{for} \quad j = 1, \dots, m.$$

We can choose δ such that the estimate (6.13) of the preceding lemma is satisfied for every λ_k with $k < k'$. Let us introduce the linear operator

$$I = \prod_{k \in A} (I_{\delta, \lambda_k})^{m_k}$$

(composition of M linear operators). It follows from the definition of $I_{\delta, \lambda}$ that I does not depend on the choice of the order of the factors I_{δ, λ_k}. Hence, by a repeated application of the preceding lemma we obtain that for every solution

$$U(t) = \sum_{k \in K} \sum_{\ell = 1}^{m_k} U_{k, \ell} F_{k, \ell}(t)$$

of (6.4), the function $V := IU$ has the form

$$V(t) = \sum_{k \in K \setminus A} \sum_{\ell = 1}^{m_k} V_{k, \ell} F_{k, \ell}(t),$$

and we have the following estimates:

$$\sum_{k \in K \setminus A} \sum_{\ell = 1}^{m_k} |U_{k, \ell}|^2 \leq c \sum_{k \in K \setminus A} \sum_{\ell = 1}^{m_k} |V_{k, \ell}|^2$$

and

$$\sum_{j=1}^{m} \int_{I_j} p_j(V(t))^2 \, dt \leq c \sum_{j=1}^{m} \int_{J_j} p_j(U(t))^2 \, dt.$$

Let us also remark that V is also a solution of (6.4). Since its initial datum V_0 belongs to $\mathcal{Z}_A$, we have

$$\sum_{k \in K \setminus A} \sum_{\ell = 1}^{m_k} |V_{k, \ell}|^2 \asymp \|V_0\|^2 \leq c \sum_{j=1}^{m} \int_{I_j} p_j(V(t))^2 \, dt$$

by the assumptions of the theorem.

Combining the last three inequalities completes the proof of the lemma.

Now we are ready to complete the proof of Theorem 6.2.

Proof of the inverse inequality in (6.7). We have to prove that

$$\|U_0\|^2 \leq c \sum_{j=1}^{m} \int_{J_j} p_j(U(t))^2 \, dt \tag{6.16}$$

for all $U_0 \in \mathcal{Z}$.

It suffices to consider the case in which λ_k has the same value, say λ, for all $k \in A$: the general case then follows by induction. Furthermore, by increasing A if necessary (recall that this only weakens the assumptions of the theorem), we may assume that λ_k has a different value for all $k \in K \backslash A$.

Set

$$V(t) = \sum_{k \in A} \sum_{\ell=1}^{m_k} U_{k,\ell} F_{k,\ell}(t), \quad V_0 = V(0), \tag{6.17}$$

and

$$W(t) = \sum_{k \in K \backslash A} \sum_{\ell=1}^{m_k} U_{k,\ell} F_{k,\ell}(t), \quad W_0 = W(0),$$

so that $U = V + W$.

Assume for the moment that

$$\|V_0\|^2 \le c \sum_{j=1}^{m} \int_{I_j} p_j(V(t))^2 \, dt. \tag{6.18}$$

Then, using the triangle inequality and then the Young inequality, we have

$$\|U_0\|^2 \le 2\|V_0\|^2 + 2\|W_0\|^2$$
$$\le c \left(\sum_{j=1}^{m} \int_{I_j} p_j(V(t))^2 \, dt \right) + 2\|W_0\|^2$$
$$\le c \left(\sum_{j=1}^{m} \int_{I_j} 2p_j(U(t))^2 + 2p_j(W(t))^2 \, dt \right) + 2\|W_0\|^2.$$

Applying the assumption (6.5) of the theorem for W instead of U, it follows that

$$\|U_0\|^2 \le c \sum_{j=1}^{m} \int_{I_j} p_j(U(t))^2 \, dt + c\|W_0\|^2.$$

Using the Riesz basis property

$$\|W_0\|^2 \le c \sum_{k \in K \backslash A} \sum_{\ell=1}^{m_k} |U_{k,\ell}|^2$$

and then applying Lemma 6.4, we conclude that

$$\|U_0\|^2 \le c \left(\sum_{j=1}^{m} \int_{I_j} p_j(U(t))^2 \, dt \right) + c \left(\sum_{j=1}^{m} \int_{J_j} p_j(U(t))^2 \, dt \right).$$

Since $I_j \subset J_j$ for every j, (6.16) follows.

It remains to prove (6.18). Since the vector space of functions of the form (6.17) is finite-dimensional, it suffices to prove that if a function of this form satisfies

$$p_j(V(t)) = 0 \quad \text{in} \quad I_j \tag{6.19}$$

for $j = 1, \ldots, m$, then $V(0) = 0$.

Observe that (6.19) implies that

$$p_j(V^{(n)}(t)) = 0 \quad \text{in the interior of} \quad I_j$$

for all $j = 1, \ldots, m$ and $n = 1, 2, \ldots$. For example, for any fixed interior point $t \in I_j$, using the continuity of the seminorm p_j, we have

$$0 \le p_j(V'(t))$$
$$= p_j\left(\lim_{h \to 0} \frac{V(t+h) - V(t)}{h}\right)$$
$$= \lim_{h \to 0} p_j\left(\frac{V(t+h) - V(t)}{h}\right)$$
$$\le \lim_{h \to 0} \frac{p_j(V(t+h)) + p_j(V(t))}{|h|},$$

and the last fraction vanishes if h is close to zero so that $t + h \in I_j$. Hence $p_j(V'(t)) = 0$.

Writing $LU = U' - \lambda U$, it follows thus from our hypothesis that

$$p_j(L^n V(t)) = 0 \quad \text{in the interior of} \quad I_j$$

for all $j = 1, \ldots, m$ and $n = 0, 1, \ldots$.

Now assume to the contrary that $V_0 \ne 0$, and let us apply this equality with the largest integer n such that $U_{k'',n} \ne 0$ for some k''. Setting $U_{k,n} = 0$ if $n > m_k$ and using the relations

$$\mathcal{A}F_{k,\ell}(t) = \lambda F_{k,\ell}(t) + F_{k,\ell-1}(t)$$

for $k \in A$ (see p. 33 and recall that $\lambda_k = \lambda$ for all $k \in A$), we have

$$L^{n-1}V(t) = \sum_{k \in A} U_{k,n} E_{k,1},$$

so that

$$p_j\left(\sum_{k \in A} U_{k,n} E_{k,1}\right) = 0, \quad j = 1, \ldots, m.$$

Applying the hypothesis of the theorem and then using the linear independence of the vectors $E_{k,1}$, we conclude that $U_{k,n} = 0$ for all k. However, this contradicts the choice of n.

Let us end this section by formulating an important consequence of Theorems 6.1 and 6.2. As in Theorem 6.2, consider again the solutions of the problem

$$U' = \mathcal{A}U, \quad U(0) = U_0, \tag{6.20}$$

in an infinite-dimensional *complex* Hilbert space $\mathcal{H}$, where $\mathcal{A}$ is an unbounded linear operator defined on some linear subspace of $\mathcal{H}$, with values in $\mathcal{H}$, satisfying hypothesis (RB) on page 34. Write $\lambda_k = i\omega_k$ and assume that

$$\text{the numbers } \omega_k \text{ are distinct.} \tag{6.21}$$

Furthermore, assume that there exist a finite subset A of K, a positive number γ', and a positive integer M such that

$$\omega_k \text{ is real and } m_k = 1 \text{ for every } k \in K \setminus A \tag{6.22}$$

and

$$\begin{cases} \text{no interval } (\omega_k - \gamma', \omega_k + \gamma') \text{ with } k \in K \setminus A \text{ contains} \\ \text{more than } M \text{ members of the family } (\omega_k)_{k \in K \setminus A}. \end{cases} \tag{6.23}$$

Finally, let $p(\cdot, \cdot)$ be a semiscalar product on $\mathcal{Z}$, and denote by $p(\cdot)$ the corresponding seminorm. Assume that

$$p(E_{k,1}) \asymp 1 \tag{6.24}$$

and

$$|p(E_{k,1}, E_{n,1})| \to 0 \quad \text{as} \quad |\omega_k - \omega_n| < \gamma' \quad \text{and} \quad k, n \to \pm\infty. \tag{6.25}$$

Theorem 6.5. *Assume* (6.21)–(6.25) *and let I be an interval of length $|I| > 2\pi/\gamma'$. Then the solutions of* (6.20) *satisfy the estimate*

$$\int_I p(U(t))^2 \, dt \asymp \|U_0\|^2 \tag{6.26}$$

for all $U_0 \in \mathcal{Z}$.

Remark. Despite the numerous hypotheses, we shall see in the next sections that this theorem has many applications.

Proof. Theorem 6.5 follows from Theorems 6.1 and 6.2 in the same way as Theorems 4.3 and 4.5 implied Theorem 4.6 in Section 4.3 (p. 67). For the proof of the uniqueness hypothesis of Theorem 6.2 we note that assumption (6.21) implies that every eigenvector E of $\mathcal{A}$ is a multiple of some element $E_{k,1}$ of the Riesz basis, say $E = cE_{k,1}$. Hence if $p(E) = 0$, then we have necessarily $c = 0$ and thus $E = 0$ because $p(E_{k,1}) \neq 0$ by assumption (6.24).

6.3 Observation of a String at Both Endpoints with Mixed Boundary Conditions

Let us return to the system studied in Sections 3.4, 4.5, and 4.7 (pp. 45, 74, 79):

$$\begin{cases} u_{tt} - u_{xx} + au = 0 & \text{in } \mathbb{R} \times (0, \ell), \\ u(t, 0) = u_x(t, \ell) = 0 & \text{for } t \in \mathbb{R}, \\ u(0, x) = u_0(x) & \text{for } x \in (0, \ell), \\ u_t(0, x) = u_1(x) & \text{for } x \in (0, \ell). \end{cases} \qquad (6.27)$$

We recall from Proposition 4.9 (p. 74) that if

$$u_0 \in V := \{v \in H^1(0, \ell) \; : \; v(0) = 0\}$$

and

$$u_1 \in H := L^2(0, \ell),$$

then (6.27) has a unique solution satisfying

$$u \in C(\mathbb{R}; V) \cap C^1(\mathbb{R}; H).$$

Furthermore, setting

$$\mu_k := \left(k - \frac{1}{2}\right) \frac{\pi}{\ell}, \quad \omega_k := \sqrt{\mu_k^2 + a},$$

and

$$E(t) := \frac{1}{2} \int_\Omega |u_x(t, x)|^2 + |u_t(t, x)|^2 \; dx,$$

we have[3]

$$u(t, x) = \sum_{k=1}^\infty (a_k e^{i\omega_k t} + a_{-k} e^{-i\omega_k t}) \sin \mu_k x$$

with suitable complex coefficients a_k and a_{-k} such that

$$E(0) \asymp \sum_{k=1}^\infty k^2 \left(|a_k|^2 + |a_{-k}|^2\right) < \infty.$$

We have postponed until now the proof of Proposition 4.13 (p. 79). Now we establish this result by applying Theorem 6.5 above. Let us recall the result we are going to prove:

Proposition 6.6. *If $T > \ell$, then the solutions of (6.27) satisfy the estimates*

$$\|u_x(\cdot, 0)\|_{L^2(0,T)}^2 + \|u(\cdot, \ell)\|_{H^1(0,T)}^2 \asymp E(0)$$

for all $u_0 \in V$ and $u_1 \in H$.

---

[3]If $\omega_k = 0$ for some k, then the term $a_k e^{i\omega_k t} + a_{-k} e^{-i\omega_k t}$ is replaced by $a_k + a_{-k} t$.

Proof. It follows from the proof of Proposition 3.4 (p. 47) that the solutions of the associated problem $U' = \mathcal{A}U$ are given by the formula

$$U(t) = \sum_{k \in \mathbb{Z}\setminus\{0\}} U_k e^{i\omega_k t} E_k$$

with

$$\omega_{-k} := -\omega_k \quad \text{and} \quad E_{\pm k} := \frac{(\sin \mu_k x, \pm i\omega_k \sin \mu_k x)}{\sqrt{\mu_k^2 + |\omega_k|^2}}$$

for $k = 1, 2, \ldots$. If $\omega_k = 0$ for some k, then the term

$$U_k e^{i\omega_k t} E_k + U_{-k} e^{-i\omega_k t} E_{-k}$$

is replaced by

$$U_k(\sin \mu_k x, 0) + U_{-k}\big((0, \sin \mu_k x) + t(\sin \mu_k x, 0)\big).$$

Since we know that

$$E(0) \asymp \|U(0)\|_{\mathcal{H}}^2,$$

the estimate of the proposition is equivalent to

$$\int_I p(U(t))^2 \, dt \asymp \|U(0)\|_{\mathcal{H}}^2,$$

where p is a Euclidean seminorm on $\mathcal{Z}$, defined by

$$p(f, g) := \sqrt{|f'(0)|^2 + |f(\ell)|^2 + |g(\ell)|^2}, \quad (f, g) \in \mathcal{Z}.$$

We will prove this inequality by applying Theorem 6.5. Since $|I| > \ell$ by assumption, there exists a number $\gamma' < 2\pi/\ell$ such that $|I| > 2\pi/\gamma'$. Using the definition of μ_k and ω_k one can readily verify that assumptions (6.21)–(6.23) are satisfied with $M = 3$ and $A = \{\pm 1, \pm 2, \ldots, \pm k'\}$ for a sufficiently large integer k'.

Condition (6.24) easily follows from the equality

$$p(E_{k,1}) = \frac{\sqrt{\mu_k^2 + 1 + |\omega_k|^2}}{\sqrt{\mu_k^2 + |\omega_k|^2}}$$

because $\mu_k^2 \asymp |\omega_k|^2 \asymp k^2$, so that the above fraction tends to 1.

For the proof of (6.25) we note first that if $|k|$ and $|n|$ are sufficiently large, then the inequality $|\omega_k - \omega_n| < \gamma'$ can occur only if k and n are consecutive integers. Hence it suffices to prove that

$$p(E_{k,1}, E_{k+1,1}) \quad \text{and} \quad p(E_{-k,1}, E_{-k-1,1})$$

tend to zero as $k \to \infty$. For this, we observe first that

$$p(E_{k,1}, E_{k+1,1}) = p(E_{-k,1}, E_{-k-1,1})$$
$$= \frac{\mu_k \mu_{k+1} + \sin \mu_k \ell \sin \mu_{k+1} \ell + \omega_k \omega_{k+1} \sin \mu_k \ell \sin \mu_{k+1} \ell}{\sqrt{\mu_k^2 + |\omega_k|^2} \sqrt{\mu_{k+1}^2 + |\omega_{k+1}|^2}}$$
$$= \frac{\mu_k \mu_{k+1} - 1 - \omega_k \omega_{k+1}}{\sqrt{\mu_k^2 + |\omega_k|^2} \sqrt{\mu_{k+1}^2 + |\omega_{k+1}|^2}}.$$

Here the denominator tends to infinity (as k^2) because $\mu_k^2 \asymp |\omega_k|^2 \asymp k^2$. Therefore, the proof will be complete if we show that the numerator remains bounded as $k \to \infty$.

Since $\mu_k \asymp k^2$ and thus

$$\omega_k - \mu_k = \frac{a}{\sqrt{\mu_k^2 + a} + \mu_k} = O(k^{-2}) = \mu_k O(k^{-4}),$$

we have

$$\mu_k \mu_{k+1} - 1 - \omega_k \omega_{k+1} = \mu_k \mu_{k+1} - 1 - \mu_k \mu_{k+1} \left(1 + O(k^{-4})\right)^2$$
$$= -1 + \mu_k \mu_{k+1} O(k^{-4})$$
$$= O(1).$$

This completes the proof.

Remark. The above proof could easily be adapted to give new proofs of Propositions 4.11 and 4.12 in Section 4.6 (p. 75), which avoid the "trick" of separating the odd and even indices.

6.4 Observation of a Coupled String–Beam System

Let us consider the one-dimensional case of the coupled system (3.12) introduced in Section 3.8 (p. 53):

$$\begin{cases} u_{tt} - u_{xx} + au + bw = 0 & \text{in} \quad \mathbb{R} \times (0, \ell), \\ w_{tt} + w_{xxxx} + cu + dw = 0 & \text{in} \quad \mathbb{R} \times (0, \ell), \\ u(t, 0) = u(t, \ell) = 0 & \text{for} \quad t \in \mathbb{R}, \\ w(t, 0) = w(t, \ell) = w_{xx}(t, 0) = w_{xx}(t, \ell) = 0 & \text{for} \quad t \in \mathbb{R}, \\ u(0, x) = u_0(x) \quad \text{and} \quad u_t(0, x) = v_0(x) & \text{for} \quad x \in (0, \ell), \\ w(0, x) = w_0(x) \quad \text{and} \quad w_t(0, x) = z_0(x) & \text{for} \quad x \in (0, \ell). \end{cases} \quad (6.28)$$

The energy of the solutions can now be written in the following form:

$$E(t) := \frac{1}{2} \int_0^\ell |u_x(t, x)|^2 + |u_t(t, x)|^2 + |w_x(t, x)|^2 + |(\Delta^{-1} w_t(t, x))_x|^2 \, dx, \quad t \in \mathbb{R}.$$

We are going to establish the following observability result:

Proposition 6.7. *For almost all quadruples* (a, b, c, d) *of real numbers, if* I *is an interval of length* $|I| > 2\ell$, *then the solutions of* (6.28) *satisfy the estimates*

$$\|u_x(\cdot, 0)\|^2_{L^2(I)} + \|w_x(\cdot, 0)\|^2_{L^2(I)} \asymp E(0) \qquad (6.29)$$

for all

$$u_0 \in H^1_0(\Omega), \quad u_1 \in L^2(\Omega), \quad w_0 \in H^1_0(\Omega), \quad and \quad w_1 \in H^{-1}(\Omega).$$

Remark. The following proof can easily be adapted to the case in which u and w are observed at different endpoints of $(0, \ell)$, or when they are observed at both endpoints. We do not insist on this here because we will prove a much more general result in the next section.

Proof. Putting

$$U = (u, v, w, z), \quad U_0 = (u_0, v_0, w_0, z_0),$$

and using the Riesz basis of Proposition 3.9 (p. 54), by Theorem 3.1 (p. 36) the solution of (6.28) is given by a series[4]

$$U(t) = \sum_{k=1}^{\infty} \sum_{j=1}^{4} U_{k,j} e^{i\omega_{k,j} t} E_{k,j}.$$

Since

$$E(0) = \frac{1}{2}\|U_0\|^2,$$

we complete the proof by verifying the conditions of Theorem 6.5 (p. 102) with the Euclidean seminorm p defined by

$$p(f, g, h, k) := \sqrt{|f'(0)|^2 + |g'(0)|^2}, \quad (f, g, h, k) \in \mathcal{Z} = Z^4.$$

Let us choose the real numbers a, b, c, d such that each 4×4 matrix $\mathcal{A}_k$ in the proof of Proposition 3.9 has four distinct eigenvalues and, moreover, such that different matrices $\mathcal{A}_k$ have no common eigenvalues. Observe that the exceptional quadruples (a, b, c, d) form a countable union of hypersurfaces[5] in $\mathbb{R}^4$, so that almost every quadruple (a, b, c, d) in $\mathbb{R}^4$ has this property.

Then condition (6.21) is satisfied. Furthermore, condition (6.22) is satisfied because all eigenvalues are simple, and because we chose real coupling coefficients. (See the last statement of Proposition 3.9.)

Taking into account the asymptotic behavior (3.15)–(3.16) of the eigenvalues and the equality

---

[4]With usual changes in a finite number of terms if generalized eigenfunctions are also present.

[5]Namely, the solutions of the algebraic equations $\omega_{k,j} = \omega_{m,n}$ for different pairs (k, j) and (m, n) in the unknowns a, b, c, d.

$$\sqrt{\gamma_k} = \mu_k := k\pi/\ell,$$

we see that condition (6.23) is satisfied with $M = 2$ for every $\gamma' < \pi/\ell$, by choosing a sufficiently large set A so as to exclude the eigenvalues $\omega_{k,j}$ with a not sufficiently large k. Since $|I| > 2\ell$, we may choose γ' so as to satisfy the inequality $|I| > 2\pi/\gamma'$, too.

In order to verify the last two conditions (6.24) and (6.25), we use the asymptotic behavior (3.13)–(3.14) of the eigenvectors $E_{k,j}$. First of all, using the equality

$$e_k(x) = \sin \mu_k x$$

we have

$$p(E_{k,j}) = \gamma_k^{-1/2} \mu_k = 1.$$

Since $\|E_{k,j}\| \asymp 1$ (because $\{E_{k,j}\}$ is a Riesz basis), this proves (6.24).

Finally, for the proof of (6.25) first observe that if $0 < |\omega_{k,j} - \omega_{m,n}| < \gamma'$ with sufficiently large indices $k \geq m$, then we necessarily have

$$0 < |\sqrt{\gamma_k} - \gamma_m| < \gamma' \qquad .$$

and $n = j + 2$ by the asymptotic formulae (3.15)–(3.16). Then using also the asymptotic relations (3.13)–(3.14) we obtain that

$$p(E_{k,j}, E_{m,n}) = (\gamma_k^{-1/2} \mu_k)o(1) + o(1)(\gamma_k^{-1/2} \mu_k) = o(1).$$

This proves (6.25) and thus completes the proof of the proposition.

Example. One may wonder whether there are exceptional parameters indeed for the validity of (6.29). The answer is yes. For instance, if we choose $\ell = \pi$ for simplicity and put

$$a = -56 + m^2, \quad b = -52, \quad c = 800, \quad d = 784 + m^2,$$

where m is an arbitrarily fixed real number, then the following formulae define a nonzero solution of (3.12) for which, however, $u_x(t,0) = w_x(t,0) = 0$ for all $t \in \mathbb{R}$:

$$u(t,x) = \cos mt(65 \sin 2x - 52 \sin 4x + 13 \sin 6x),$$
$$w(t,x) = \cos mt(-65 \sin 2x + 40 \sin 4x - 5 \sin 6x).$$

See [76] for a more general example.

6.5 Observation of a Coupled System: A General Result

In this section we generalize the result of the preceding one to all spatial dimensions and to more general observations. Let us consider the system of Section 3.8 (p. 53):

$$\begin{cases} u'' - \Delta u + au + bw = 0 & \text{in } \mathbb{R} \times \Omega, \\ w'' + \Delta^2 w + cu + dw = 0 & \text{in } \mathbb{R} \times \Omega, \\ u = 0 & \text{on } \mathbb{R} \times \Gamma, \\ w = \Delta w = 0 & \text{on } \mathbb{R} \times \Gamma, \\ u(0) = u_0 \text{ and } u'(0) = u_1 & \text{in } \Omega, \\ w(0) = w_0 \text{ and } w'(0) = w_1 & \text{in } \Omega. \end{cases} \tag{6.30}$$

We recall from Proposition 3.8 (p. 53) that the problem is well-posed in the Hilbert space

$$\mathcal{H} := H_0^1(\Omega) \times H_0^1(\Omega) \times L^2(\Omega) \times H^{-1}(\Omega).$$

We are interested in the validity of the estimates

$$\int_{J_u} \int_{\Gamma_u} |\partial_\nu u|^2 \, d\Gamma \, dt + \int_{J_w} \int_{\Gamma_w} |\partial_\nu w|^2 \, d\Gamma \, dt \asymp E(0), \tag{6.31}$$

where J_u, J_w are two given intervals, Γ_u, Γ_w are two open subsets of Γ, and the energy of the solutions is defined by the formula[6]

$$E(t) := \frac{1}{2} \int_\Omega |\nabla u(t,x)|^2 + |u'(t,x)|^2 + |\nabla w(t,x)|^2 + |\nabla \Delta^{-1} w'(t,x)|^2 \, dx, \quad t \in \mathbb{R}.$$

We need a geometric condition. Assume that there exist two points $x_u, x_w \in \mathbb{R}^N$ such that (see Figure 6.1 and the first remark below)

$$(x - x_u) \cdot \nu(x) \leq 0 \quad \text{for all} \quad x \in \Gamma \backslash \Gamma_u \tag{6.32}$$

and

$$(x - x_w) \cdot \nu(x) \leq 0 \quad \text{for all} \quad x \in \Gamma \backslash \Gamma_w. \tag{6.33}$$

Then putting

$$R = R_u := \sup_{x \in \Omega} |x - x_u|, \tag{6.34}$$

we have the following result:

Proposition 6.8. *Assume* (6.32) *and* (6.33). *Fix an interval J_u of length $|J_u| > 2R$ and an arbitrary interval J_w. Then the estimates* (6.31) *hold for almost all quadruples* (a, b, c, d) *of complex numbers.*

---

[6]As in Section 3.5 (p. 48), Δ^{-1} denotes the inverse of the restriction of Δ to $H_0^1(\Omega)$.

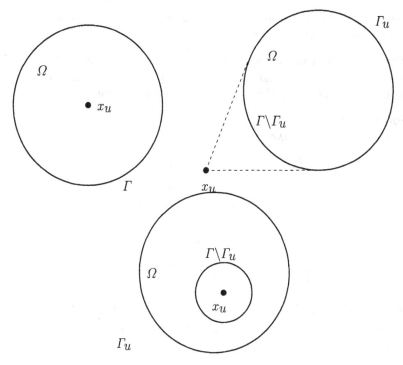

Fig. 6.1. The geometric condition

Remarks.

- Intuitively, (6.32) and (6.33) require that both Γ_u and Γ_w represent more than half of the boundary (but not an arbitrary half!). They are obviously satisfied if $\Gamma_u = \Gamma_w = \Gamma$. They are also satisfied if Ω is an annular region and $\Gamma_u = \Gamma_w$ is its outer boundary. In the one-dimensional case $\Omega = (0, \ell)$ we may take

$$\Gamma_u = \{0\}, \Gamma_u = \{\ell\}, \quad \text{or} \quad \Gamma_u = \{0, \ell\},$$

and independently

$$\Gamma_w = \{0\}, \Gamma_w = \{\ell\}, \quad \text{or} \quad \Gamma_w = \{0, \ell\}.$$

- Notice that R is the smallest real number for which the open ball $B_R(x_u)$ of center x_u and radius R contains the domain Ω.
- The proposition and its proof remain valid if we replace (6.32)–(6.34) by the *geometric control condition* of Bardos, Lebeau, and Rauch [11].
- The proposition was first established in [76] for the case in which Ω is a ball, by the method of the preceding section, and then in [82] for the general case, in an indirect way. The constructive proof given below is taken from [80], where systems of more than two equations were also studied. Recently, Mehrenberger [107] obtained rather precise information on the

structure and size of the set of exceptional parameters in case of several coupled equations.

For the proof of the proposition we shall use the direct and inverse inequalities for the wave equation and for the Petrovsky system in a general domain. Let us first recall these results for the wave equation.

Proposition 6.9.

(a) *The solutions of the problem*

$$\begin{cases} u'' - \Delta u + f = 0 & in \quad \mathbb{R} \times \Omega, \\ u = 0 & on \quad \mathbb{R} \times \Gamma, \\ u(0) = u'(0) = 0 & in \quad \Omega, \end{cases} \tag{6.35}$$

satisfy for every $T > 0$ the direct inequality

$$\int_0^T \int_\Gamma |\partial_\nu u|^2 \, d\Gamma \, dt \leq c \int_0^T \int_\Omega |f(t,x)|^2 \, dx \, dt$$

for all $f \in L^1_{\text{loc}}(\mathbb{R}; L^2(\Omega))$, with a constant c depending only on T.

(b) *Assume (6.32) and let I be an interval of length $> 2R$, where R is given by (6.34). Then the solutions of the problem*

$$\begin{cases} u'' - \Delta u = 0 & in \quad \mathbb{R} \times \Omega, \\ u = 0 & on \quad \mathbb{R} \times \Gamma, \\ u(0) = u_0 \quad and \quad u'(0) = u_1 & in \quad \Omega, \end{cases} \tag{6.36}$$

satisfy the estimates

$$\int_I \int_{\Gamma_u} |\partial_\nu u|^2 \, d\Gamma \, dt \asymp \int_\Omega |\nabla u_0|^2 + |u_1|^2 \, dx.$$

Remarks.

- Part (a) of the proposition is due to Lasiecka and Triggiani [91]. (The well-posedness of this inhomogeneous problem follows from the remark after Theorem 2.1, p. 10.) Part (b) was first obtained by Ho [50], under a stronger assumption on the length of I. His result was improved by Lions [96] by an indirect argument based on Carleman's uniqueness theorem. Subsequently, a simple constructive proof was given in [62] that avoided the use of Carleman's result. Although these proofs are based on the multiplier method, which is not the subject of this book, for the convenience of the reader we reproduce in the next, optional, section the proof of Proposition 6.9 for *star-shaped* domains, and we refer to Lions [97] or to [67] for the general case. The condition $T > 2R$ is optimal if Ω is a ball centered at x_u, or more generally if Ω contains a line segment of length $2R$. This can be shown by a simple argument demonstrating the finite propagation speed for the wave equation; see, e.g., [67], Remark 3.6.

- More complete results were established by Bardos, Lebeau, and Rauch [11], using microlocal analysis.
- If we replace the Dirichlet boundary condition by the Neumann boundary condition in (6.35) and (6.36), then the just-mentioned methods provide less precise results; see [97] or [67] again. We will show in Chapter 7 (see Proposition 7.1, p. 128) that except for the one-dimensional case, the corresponding natural boundary integral can no longer be expressed in terms of Sobolev norms of the initial data.

Next we recall some analogous results on the Petrovsky system.

Proposition 6.10.
(a) The solutions of the problem

$$\begin{cases} w'' + \Delta^2 w + f = 0 & \text{in } \mathbb{R} \times \Omega, \\ w = \Delta w = 0 & \text{on } \mathbb{R} \times \Gamma, \\ w(0) = w'(0) = 0 & \text{in } \Omega, \end{cases} \tag{6.37}$$

satisfy for every $T > 0$ the direct inequality

$$\int_0^T \int_\Gamma |\partial_\nu w|^2 \, d\Gamma \, dt \le c \int_0^T \|f\|_{H^{-1}(\Omega)}^2 \, dt$$

for all $f \in L^1_{\text{loc}}(\mathbb{R}; H^{-1}(\Omega))$, with a constant c depending only on T.

(b) Assume (6.33) and let I be an interval of arbitrary length. Then the solutions of the problem

$$\begin{cases} w'' + \Delta^2 w = 0 & \text{in } \mathbb{R} \times \Omega, \\ w = \Delta w = 0 & \text{on } \mathbb{R} \times \Gamma, \\ w(0) = w_0 \quad \text{and} \quad w'(0) = w_1 & \text{in } \Omega, \end{cases} \tag{6.38}$$

satisfy the estimates

$$\int_I \int_{\Gamma_w} |\partial_\nu w|^2 \, d\Gamma \, dt \asymp \|w_0\|_{H_0^1(\Omega)}^2 + \|w_1\|_{H^{-1}(\Omega)}^2.$$

Remarks.

- Part (a) of the proposition is due to Lions [96]. He also established part (b) for sufficiently long intervals I. His assumption on the length of I was weakened by an elementary argument in [62] and then completely relaxed by Zuazua [139], who applied an indirect argument based on Holmgren's uniqueness theorem. An alternative, constructive, proof was later given in [63], without using Holmgren's theorem. For the convenience of the reader, in the optional Section 6.7 we reproduce this last proof for *star-shaped* domains, and we refer to [67] for the general case. Let us also note that using microlocal analysis, more general results were established later by Lebeau [92].

- As in the case of the wave equation, the situation is more complex in the case of other boundary conditions. We shall examine this question in Chapter 7 (see Proposition 7.11, p. 143).

Turning to the proof of Proposition 6.8, let us rewrite the problem (6.30) in the form

$$U' = \mathcal{A}U, \quad U(0) = U_0,$$

as in Section 3.8 (p. 53), and let us introduce the same Riesz basis as in Proposition 3.9.

Given two intervals J_u and J_w with $|J_u| > 2R$, fix two new intervals I_u and I_w such that $|I_u| > 2R$, the closure of I_u belongs to the interior of J_u, and the closure of I_w belongs to the interior of J_w. We are going to apply Theorem 6.2 (p. 93) with $m = 2$, $I_1 = I_u$, $I_2 = I_w$, and with the seminorms given by

$$p_1(u, w, h, k) := \|\partial_\nu u\|_{L^2(\Gamma_u)} \quad \text{and} \quad p_2(u, w, h, k) := \|\partial_\nu w\|_{L^2(\Gamma_w)}.$$

We have to show that for a sufficiently large integer k', the estimates

$$\int_{I_u} \int_{\Gamma_u} |\partial_\nu u|^2 \, d\Gamma \, dt + \int_{I_w} \int_{\Gamma_w} |\partial_\nu w|^2 \, d\Gamma \, dt \asymp E(0) \qquad (6.39)$$

hold for all finite sums of the form

$$U(t) = \sum_{k=k'}^{\infty} \sum_{j=1}^{4} U_{k,j} e^{i\omega_{k,j} t} E_{k,j}.$$

(By rearranging the terms if necessary, we may assume that $\gamma_k \geq \gamma_{k'}$ for all $k \geq k'$.)

Furthermore, we have to show that if $\mathcal{A}U_0 = \lambda U_0$ for some $U_0 = (u, w, h, k) \in \mathcal{Z}$ and if

$$\partial_\nu u = 0 \quad \text{on} \quad \Gamma_u \quad \text{and} \quad \partial_\nu w = 0 \quad \text{on} \quad \Gamma_w,$$

then

$$u = w = h = k = 0 \quad \text{in} \quad \Omega.$$

Fix a large integer k', to be chosen later. Let us write the solution of (6.30) in the form

$$u = u_f + \widehat{u} \quad \text{and} \quad w = w_g + \widehat{w},$$

where u_f solves (6.35) with $f := au + bw$, $\widehat{u}$ solves (6.36), w_g solves (6.37) with $g := cu + dw$, and $\widehat{w}$ solves (6.38). Applying part (a) of the preceding two propositions for u_f and w_g, we obtain that

$$\int_{I_u} \int_{\Gamma_u} |\partial_\nu u_f|^2 \, d\Gamma \, dt + \int_{I_w} \int_{\Gamma_w} |\partial_\nu w_g|^2 \, d\Gamma \, dt$$

$$\leq c \int_{I_u} \|u\|_{L^2(\Omega)}^2 + \|w\|_{L^2(\Omega)}^2 \, dt + c \int_{I_w} \|u\|_{H^{-1}(\Omega)}^2 + \|w\|_{H^{-1}(\Omega)}^2 \, dt.$$

Choosing an interval I' containing both I_u and I_w, using the continuity of the embedding $L^2(\Omega) \subset H^{-1}(\Omega)$ and then the definition of the eigenvalue $\gamma_{k'}$, it follows that

$$\int_{I_u} \int_{\Gamma_u} |\partial_\nu u_f|^2 \, d\Gamma \, dt + \int_{I_w} \int_{\Gamma_w} |\partial_\nu w_g|^2 \, d\Gamma \, dt \leq \frac{c}{\gamma_{k'}} \int_{I'} \|u\|^2_{H_0^1(\Omega)} + \|w\|^2_{H_0^1(\Omega)} \, dt.$$

Now using the well-posedness estimate (3.5) of Theorem 3.1 (p. 36), the definition of the energy, and the equality

$$E(0) = \frac{1}{2}\|U_0\|^2,$$

we conclude that

$$\int_{I_u} \int_{\Gamma_u} |\partial_\nu u_f|^2 \, d\Gamma \, dt + \int_{I_w} \int_{\Gamma_w} |\partial_\nu w_g|^2 \, d\Gamma \, dt \leq \frac{c_1}{\gamma_{k'}} E(0) \tag{6.40}$$

with some constant c_1.

Next, applying part (b) of the preceding two propositions for $\widehat{u}$ and $\widehat{w}$, we obtain that

$$c_2 E(0) \leq \int_{I_u} \int_{\Gamma_u} |\partial_\nu \widehat{u}|^2 \, d\Gamma \, dt + \int_{I_w} \int_{\Gamma_w} |\partial_\nu \widehat{w}|^2 \, d\Gamma \, dt \leq c_3 E(0) \tag{6.41}$$

with two positive constants c_2 and c_3.

Applying the Young inequalities

$$|\partial_\nu u|^2 \leq 2|\partial_\nu u_f|^2 + 2|\partial_\nu \widehat{u}|^2 \quad \text{and} \quad |\partial_\nu w|^2 \leq 2|\partial_\nu w_g|^2 + 2|\partial_\nu \widehat{w}|^2,$$

we conclude from (6.40) and (6.41) the first half of (6.39):

$$\int_{I_u} \int_{\Gamma_u} |\partial_\nu u|^2 \, d\Gamma \, dt + \int_{I_w} \int_{\Gamma_w} |\partial_\nu w|^2 \, d\Gamma \, dt \leq \left(\frac{2c_1}{\gamma_{k'}} + 2c_3 \right) E(0).$$

Furthermore, using the Young inequalities in the form

$$|\partial_\nu \widehat{u}|^2 \leq 2|\partial_\nu u_f|^2 + 2|\partial_\nu u|^2 \quad \text{and} \quad |\partial_\nu \widehat{w}|^2 \leq 2|\partial_\nu w_g|^2 + 2|\partial_\nu w|^2,$$

we also conclude from (6.40) and (6.41) that

$$\left(\frac{c_2}{2} - \frac{c_1}{\gamma_{k'}} \right) E(0) \leq \int_{I_u} \int_{\Gamma_u} |\partial_\nu u|^2 \, d\Gamma \, dt + \int_{I_w} \int_{\Gamma_w} |\partial_\nu w|^2 \, d\Gamma \, dt.$$

Therefore, if we choose k' so large at the beginning that

$$\frac{c_2}{2} - \frac{c_1}{\gamma_{k'}} > 0,$$

then (6.39) follows.

It remains to establish the above-mentioned uniqueness property. For this, let us choose the coupling coefficients a, b, c, and d such that

$$\omega_{k,j} \neq \omega_{m,n} \quad \text{whenever} \quad \gamma_k \neq \gamma_m.$$

As in the preceding section, the exceptional quadruples (a, b, c, d) form a countable union of hypersurfaces, so that almost all quadruples (a, b, c, d) have the required property.

Now assume that $\mathcal{A}U_0 = \lambda U_0$ for some $U_0 = (u, w. h, k) \in \mathcal{Z}$,

$$\partial_\nu u = 0 \quad \text{on} \quad \Gamma_u, \quad \text{and} \quad \partial_\nu w = 0 \quad \text{on} \quad \Gamma_w.$$

Then u, w, h, k are scalar multiples of a common eigenfunction of $-\Delta$ in Ω with Dirichlet boundary condition. Using Carleman's unique continuation theorem[7] it follows that $u = w = 0$ in Ω. Since the relation $\mathcal{A}U_0 = \lambda U_0$ implies that $h = \lambda u$ and $k = \lambda w$, we also have $h = k = 0$ in Ω. Thus $U_0 = 0$, and the proof is complete.

6.6 * Proof of Proposition 6.9 by the Multiplier Method

By considering the real and imaginary parts of the solutions, it is suffucent to establish the desired estimates in the real case. Furthermore, by a density argument it suffices to consider solutions of (6.35) and (6.36) (p. 110), or more generally of the problem

$$\begin{cases} u'' - \Delta u + f = 0 & \text{in} \quad \mathbb{R} \times \Omega, \\ u = 0 & \text{on} \quad \mathbb{R} \times \Gamma, \\ u(0) = u_0 \quad \text{and} \quad u'(0) = u_1 & \text{in} \quad \Omega, \end{cases} \qquad (6.42)$$

with

$$u_0 \in H^2(\Omega) \cap H_0^1(\Omega), \quad u_1 \in H_0^1(\Omega), \quad \text{and} \quad f \in C^1(\mathbb{R}; L^2(\Omega)).$$

Then the solution satisfies

$$u \in C(\mathbb{R}; H^2(\Omega)) \cap C^1(\mathbb{R}; H^1(\Omega)) \cap C^2(\mathbb{R}; L^2(\Omega))$$

in view of the remark after Theorem 2.1 on p. 10, and this regularity property justifies all computations that follow. In the sequel all solutions are assumed to have this regularity.

For the sake of simplicity, we consider only the case of *convex* or more generally *star-shaped* domains; i.e., we assume that there exists a point $x_u \in \Omega$ such that

$$(x - x_u) \cdot \nu(x) > 0 \quad \text{for all} \quad x \in \Gamma. \qquad (6.43)$$

---

[7]A simple proof of this theorem was given by Garofalo and Lin in [37], [38].

This implies (6.32) with $\Gamma_u = \Gamma$. The general case can be proved by a slight adaptation of the arguments below, as explained later in this section.

The proof is based on the *multiplier method*. Our main tool is the following technical lemma, which goes back essentially at least to Rellich [119]. Let us introduce the notation

$$m(x) := x - x_u \quad \text{and} \quad Mu := 2m \cdot \nabla u + (N-1)u,$$

where the dot stands for the usual scalar product in $\mathbb{R}^N$.

Lemma 6.11. *The solutions of* (6.42) *satisfy for every* $T > 0$ *the following identity:*

$$\int_0^T \int_\Gamma (m \cdot \nu)(\partial_\nu u)^2 \, d\Gamma \, dt$$

$$= \left[\int_\Omega u' Mu \, dx \right]_0^T + \int_0^T \int_\Omega (u')^2 + |\nabla u|^2 + f Mu \, dx \, dt. \quad (6.44)$$

Proof. Integrating by parts, we obtain that

$$- \int_0^T \int_\Omega f Mu \, dx \, dt = \int_0^T \int_\Omega (u'' - \Delta u) Mu \, dx \, dt \quad (6.45)$$

$$= \left[\int_\Omega u' Mu \, dx \right]_0^T - \int_0^T \int_\Gamma (\partial_\nu u) Mu \, d\Gamma \, dt$$

$$+ \int_0^T \int_\Omega -u' Mu' + \nabla u \cdot \nabla (Mu) \, dx \, dt.$$

Let us transform the last integral. We have

$$u' Mu' = 2u'm \cdot \nabla u' + (N-1)(u')^2 = m \cdot \nabla(u')^2 + (N-1)(u')^2,$$

so that integrating by parts and using the relation div $m = N$, we obtain the equality

$$- \int_\Omega u' Mu' \, dx = - \int_\Omega m \cdot \nabla(u')^2 + (N-1)(u')^2 \, dx \quad (6.46)$$

$$= - \int_\Gamma (m \cdot \nu)(u')^2 \, d\Gamma + \int_\Omega (u')^2 \, dx.$$

Next, applying the summation convention of repeated indices and using the relation $\partial_i m_k = \delta_{ik}$, we have

$$\nabla u \cdot \nabla(Mu) = (\partial_i u)\partial_i \big(2m_k \partial_k u + (N-1)u\big)$$

$$= 2(\partial_i u)(\partial_i m_k)(\partial_k u) + 2m_k(\partial_i u)(\partial_k \partial_i u) + (N-1)|\nabla u|^2$$

$$= m \cdot \nabla(|\nabla u|^2) + (N+1)|\nabla u|^2.$$

Hence

$$\int_\Omega \nabla u \cdot \nabla(Mu) \, dx = \int_\Omega m \cdot \nabla(|\nabla u|^2) + (N+1)|\nabla u|^2 \, dx \qquad (6.47)$$

$$= \int_\Gamma (m \cdot \nu)|\nabla u|^2 \, d\Gamma + \int_\Omega |\nabla u|^2 \, dx.$$

Substituting (6.46) and (6.47) into (6.45), we conclude that

$$-\int_0^T \int_\Omega f Mu \, dx \, dt = \left[\int_\Omega u' Mu \, dx \right]_0^T + \int_0^T \int_\Omega (u')^2 + |\nabla u|^2 \, dx \, dt$$

$$- \int_0^T \int_\Gamma (\partial_\nu u) Mu + (m \cdot \nu)((u')^2 - |\nabla u|^2) \, d\Gamma \, dt.$$

Until now we did not use the boundary condition in (6.42). Now we observe that since $u = 0$ on the boundary, we also have

$$u' = 0, \quad \nabla u = (\partial_\nu u)\nu, \quad \text{and} \quad Mu = 2(m \cdot \nu)\partial_\nu u \quad \text{on} \quad \Gamma.$$

Hence the expression in the boundary integral reduces to $(m \cdot \nu)(\partial_\nu u)^2$, and (6.44) follows.

Next we recall from [62] the following lemma:

Lemma 6.12. *Given $u \in H^1(\Omega)$ arbitrarily, we have the following identity:*

$$\int_\Omega (Mu)^2 \, dx = \int_\Omega |2m \cdot \nabla u|^2 + (1 - N^2)u^2 \, dx + (2N - 2) \int_\Gamma (m \cdot \nu)u^2 \, d\Gamma.$$

Hence, if u also vanishes on Γ, then we have

$$\int_\Omega (Mu)^2 \, dx \le 4R^2 \int_\Omega |\nabla u|^2 \, dx.$$

Proof. We integrate by parts, and we use again the relation div $m \equiv n$ as follows:

$$\int_\Omega (Mu)^2 \, dx = \int_\Omega |2m \cdot \nabla u + (N - 1)u|^2 \, dx$$

$$= \int_\Omega |2m \cdot \nabla u|^2 + (N - 1)^2 u^2 + 4(N - 1)u(m \cdot \nabla u) \, dx$$

$$= \int_\Omega |2m \cdot \nabla u|^2 + (N - 1)^2 u^2 + (2N - 2)m \cdot \nabla(u^2) \, dx$$

$$= \int_\Omega |2m \cdot \nabla u|^2 + (N - 1)^2 u^2 - N(2N - 2)u^2 \, dx$$

$$+ (2N - 2) \int_\Gamma (m \cdot \nu)u^2 \, d\Gamma.$$

We conclude by remarking that

$$(N - 1)^2 - N(2N - 2) = 1 - N^2.$$

Now we prove the direct inequality in star-shaped domains.

Proof of part (a) and of the direct inequality in part (b) of Proposition 6.9. The continuous factor $m \cdot \nu$ has a strictly positive minimum on the compact boundary Γ by our geometric assumption (6.43). Hence the left-hand side of the identity (6.44) is minorized by a positive constant multiple of

$$\int_0^T \int_\Gamma (\partial_\nu u)^2 \, d\Gamma \, dt.$$

Furthermore, putting

$$E := \frac{1}{2} \int_\Omega (u')^2 + (\nabla u)^2 \, dx$$

as usual and applying the preceding lemma, we have

$$\left| \int_\Omega u' M u \, dx \right| \le \int_\Omega R(u')^2 + \frac{1}{4R}(Mu)^2 \, dx \le R \int_\Omega (u')^2 + |\nabla u|^2 \, dx = 2RE.$$

Therefore, using the preceding lemma again, the right-hand side of the identity (6.44) is majorized by

$$2RE(0) + 2RE(T) + 2 \int_0^T E(t) \, dt + 2R \int_0^T \|f(t)\|_{L^2(\Omega)} \sqrt{2E(t)} \, dt$$

and hence by

$$(4R + 2T)\|E\|_{L^\infty(0,T)} + 2R\|f\|_{L^1(0,T;L^2(\Omega))} \|2E\|_{L^\infty(0,T)}^{1/2}.$$

If $f = 0$, then the energy E is conserved because we are in the skew-adjoint case, so that the last expression reduces to

$$(4R + 2T)E(0) = (T + 2R)\left(\|u_0\|_{H_0^1(\Omega)}^2 + \|u_1\|_{L^2(\Omega)}^2 \right).$$

This proves the direct inequality in part (b) of the proposition for intervals of the form $I = (0, T)$. The general case follows by using the translation invariance of the differential equation in (6.36) and the conservation of the energy.

If $u_0 = u_1 = 0$, then using the variation of constants formula for the associated abstract first-order problem with $F = (0, f)$ (see the remark following Theorem 2.1 on p. 10 and take into account that we are in the skew-adjoint case), we obtain the estimate

$$\|2E\|_{L^\infty(0,T)}^{1/2} \le \|f\|_{L^1(0,T;L^2(\Omega))}.$$

Therefore, the last expression is majorized by

$$(6R + 2T)\|f\|_{L^1(0,T;L^2(\Omega))}^2$$

as required.

Remark. It is easy to adapt the above proof to general domains: it suffices to replace the vector field m of Lemma 6.11 by an arbitrary sufficiently smooth vector field satisfying $m = \nu$ on the boundary Γ. Such vector fields can be constructed by using a partition of unity. See Lions [97] or [67] for the details.

Proof of the inverse inequality in part (b) of Proposition 6.9. As above, it suffices to consider intervals of the form $I = (0, T)$. The right-hand side of the identity (6.44), since the term fMu is missing, is *minorized* by

$$(2T - 4R)E(0) = (T - 2R)\big(\|u_0\|^2_{H_0^1(\Omega)} + \|u_1\|^2_{L^2(\Omega)}\big).$$

Since $T - 2R > 0$ by assumption, we conclude by remarking that the left-hand side of the identity (6.44) is *majorized* by

$$R \int_0^T \int_\Gamma (\partial_\nu u)^2 \, d\Gamma \, dt$$

because $m \cdot \nu \leq R$ on the boundary.

Remark. Let us emphasize that we did not use the star-shapedness of Ω in the proof of the inverse inequality. In the general case the left-hand side of the identity (6.44) is majorized by

$$R \int_0^T \int_{\Gamma_u} (\partial_\nu u)^2 \, d\Gamma \, dt.$$

6.7 * Proof of Proposition 6.10 by the Multiplier Method

For the sake of simplicity we restrict ourselves again to the case of *star-shaped* domains; i.e., we assume that there exists a point $x_w \in \Omega$ such that

$$(x - x_w) \cdot \nu(x) > 0 \quad \text{for all} \quad x \in \Gamma. \tag{6.48}$$

This implies (6.33) with $\Gamma_w = \Gamma$. The general case can be proved by the same type of adaptation as in the preceding section.

If we replace the solution w by $\Delta^{-1}w$ and f by $\Delta^{-1}f$, then Proposition 6.10 may be reformulated in the following equivalent form:

Proposition 6.13. *Assume* (6.48).

(a) *If* $f \in L^1_{\text{loc}}(\mathbb{R}; H_0^1(\Omega))$, *then the solution of the problem*

$$\begin{cases} w'' + \Delta^2 w + f = 0 & \text{in} \quad \mathbb{R} \times \Omega, \\ w = \Delta w = 0 & \text{on} \quad \mathbb{R} \times \Gamma, \\ w(0) = w'(0) = 0 & \text{in} \quad \Omega, \end{cases} \tag{6.49}$$

satisfies for every $T > 0$ *the direct inequality*

$$\int_0^T \int_\Gamma |\partial_\nu \Delta w|^2 \, d\Gamma \, dt \le c \int_0^T \|f\|_{H_0^1(\Omega)}^2 \, dt$$

with a constant c depending only on T.

(b) Let J be an arbitrary interval. If

$$w_0 \in V := \{v \in H^3(\Omega) \; : \; v = \Delta v = 0 \quad \text{on} \quad \Gamma\} \quad \text{and} \quad w_1 \in H_0^1(\Omega),$$

then the solution of the problem

$$\begin{cases} w'' + \Delta^2 w = 0 & \text{in} \quad \mathbb{R} \times \Omega, \\ w = \Delta w = 0 & \text{on} \quad \mathbb{R} \times \Gamma, \\ w(0) = w_0 \quad \text{and} \quad w'(0) = w_1 & \text{in} \quad \Omega, \end{cases} \tag{6.50}$$

satisfies the estimates

$$\int_J \int_\Gamma |\partial_\nu \Delta w|^2 \, d\Gamma \, dt \asymp \int_\Omega |\nabla \Delta w_0|^2 + |\nabla w_1|^2 \, dx.$$

By a density argument it suffices to consider solutions of (6.49) and (6.50), or more generally of the problem

$$\begin{cases} w'' + \Delta^2 w + f = 0 & \text{in} \quad \mathbb{R} \times \Omega, \\ w = \Delta w = 0 & \text{on} \quad \mathbb{R} \times \Gamma, \\ w(0) = w_0 \quad \text{and} \quad w'(0) = w_1 & \text{in} \quad \Omega, \end{cases} \tag{6.51}$$

with $w_0, w_1 \in Z$ and a continuously differentiable function $f : \mathbb{R} \to H_0^1(\Omega)$. Then the solution satisfies

$$w \in C(\mathbb{R}; H^5(\Omega)) \cap C^1(\mathbb{R}; H^3(\Omega)) \cap C^2(\mathbb{R}; H^1(\Omega)),$$

which justifies the computations that follow. In the rest of this section all solutions are assumed to have this regularity.

Finally, as a consequence of the skew-adjoint character of (6.50) and of its translation invariance, it suffices to consider intervals of the form $J = (0, T)$ in part (b), too.

Setting this time

$$m(x) = x - x_w \quad \text{and} \quad Mw := 2m \cdot \nabla \Delta w + n \Delta w,$$

we first establish the following Rellich-type identity:

Lemma 6.14. *The real-valued solutions of the problem* (6.51) *satisfy for every* $T > 0$ *the following identity:*

$$\int_0^T \int_\Gamma (m \cdot \nu)(\partial_\nu \Delta w)^2 + (m \cdot \nu)(\partial_\nu w')^2 \, d\Gamma \, dt$$

$$= \left[-\int_\Omega w' M w \, dx \right]_0^T + \int_0^T \int_\Omega 2|\nabla \Delta w|^2 + 2(\Delta w')^2 - f M w \, dx \, dt. \tag{6.52}$$

Proof. Integrating by parts, we obtain that

$$-\int_0^T \int_\Omega f M w \, dx \, dt = \int_0^T \int_\Omega (w'' + \Delta^2 w) M w \, dx \, dt \tag{6.53}$$

$$= \left[\int_\Omega w' M w \, dx\right]_0^T + \int_0^T \int_\Gamma (\partial_\nu \Delta w) M w \, d\Gamma \, dt$$

$$- \int_0^T \int_\Omega w' M w' + \nabla \Delta w \cdot \nabla(M w) \, dx \, dt.$$

Let us transform the last integral. Using the relations div $m = N$ and $\partial_i m_k = \delta_{ik}$, we have

$$-\int_\Omega w' M w' \, dx$$

$$= -\int_\Omega 2 w' m_i \partial_i \partial_j^2 w' + n w' \Delta w' \, dx$$

$$= \int_\Omega 2(\partial_j w') m_i (\partial_i \partial_j w') + 2 w' (\partial_j m_i)(\partial_i \partial_j w') - n w' \Delta w' \, dx$$

$$- \int_\Gamma 2 w' m_i \nu_j (\partial_i \partial_j w') \, d\Gamma$$

$$= \int_\Omega m \cdot \nabla(|\nabla w'|^2) + (2 - n) w' \Delta w' \, dx - \int_\Gamma 2 w' m_i \nu_j (\partial_i \partial_j w') \, d\Gamma$$

$$= \int_\Omega -n|\nabla w'|^2 + (n - 2)|\nabla w'|^2 \, dx$$

$$+ \int_\Gamma -2 w' m_i \nu_j (\partial_i \partial_j w') + (m \cdot \nu)|\nabla w'|^2 + (2 - n) w' \partial_\nu w' \, d\Gamma.$$

Since $w = 0$ and hence $w' = 0$ on the boundary, $\nabla w' = (\partial_\nu w')\nu$ on Γ. Therefore, the first and last terms in the boundary integral vanish, while the second is equal to $(m \cdot \nu)|\partial_\nu w'|^2$. Hence we conclude that

$$-\int_\Omega w' M w' \, dx = -2\int_\Omega |\nabla w'|^2 \, dx + \int_\Gamma (m \cdot \nu)(\partial_\nu w')^2 \, d\Gamma. \tag{6.54}$$

Next we have

$$-\int_\Omega \nabla \Delta w \cdot \nabla(M w) \, dx$$

$$= -\int_\Omega (\partial_i \Delta w) \partial_i (2 m_k \partial_k \Delta w + n \Delta w) \, dx$$

$$= -\int_\Omega 2(\partial_i \Delta w)(\partial_i m_k)(\partial_k \Delta w) + 2 m_k (\partial_i \Delta w)(\partial_k \partial_i \Delta w) + n|\nabla \Delta w|^2 \, dx$$

$$= -\int_\Omega m \cdot \nabla(|\nabla \Delta w|^2) + (n + 2)|\nabla \Delta w|^2 \, dx$$

$$= -2\int_\Omega |\nabla \Delta w|^2 \, dx - \int_\Gamma (m \cdot \nu)|\nabla \Delta w|^2 \, d\Gamma.$$

Since $\Delta w = 0$ and therefore $\nabla \Delta w = (\partial_\nu \Delta w)\nu$ on the boundary, we conclude that

$$-\int_\Omega \nabla \Delta w \cdot \nabla(Mw)\ dx = -2\int_\Omega |\nabla \Delta w|^2\ dx - \int_\Gamma (m \cdot \nu)(\partial_\nu \Delta w)^2\ d\Gamma. \quad (6.55)$$

Substituting (6.54) and (6.55) into (6.53), we obtain the following equality:

$$-\int_0^T \int_\Omega fMw\ dx\ dt$$
$$= \left[\int_\Omega w'Mw\ dx\right]_0^T - 2\int_0^T \int_\Omega |\nabla \Delta w|^2 + |\nabla w'|^2\ dx\ dt$$
$$+ \int_0^T \int_\Gamma (\partial_\nu \Delta w)Mw + (m \cdot \nu)(\partial_\nu w')^2 - (m \cdot \nu)(\partial_\nu \Delta w)^2\ d\Gamma\ dt.$$

Using again the relations $\Delta w = 0$ and $\nabla \Delta w = (\partial_\nu \Delta w)\nu$ on the boundary, we have $\nabla \Delta w = (\partial_\nu \Delta w)\nu$ and therefore

$$M\Delta w = 2(m \cdot \nu)\partial_\nu \Delta w \quad \text{on} \quad \Gamma.$$

Hence the first term in the boundary integral is equal to $2(m \cdot \nu)(\partial_\nu \Delta w)^2$, and (6.52) follows.

Next we recall from [62] the following lemma:

Lemma 6.15. *Every real function $w \in H^3(\Omega)$ satisfies the following identity:*

$$\int_\Omega (Mw)^2\ dx = \int_\Omega |2m \cdot \nabla \Delta w|^2 - N^2(\Delta w)^2\ dx + 2N\int_\Gamma (m \cdot \nu)(\Delta w)^2\ d\Gamma.$$

Hence if Δw also vanishes on Γ, then we have

$$\int_\Omega (Mw)^2\ dx \leq 4R^2 \int_\Omega |\nabla \Delta w|^2\ dx.$$

Proof. Set $u := \Delta w$ for brevity. Integrating by parts, we obtain the following equality:

$$\int_\Omega (Mu)^2\ dx = \int_\Omega |2m \cdot \nabla u + Nu|^2\ dx$$
$$= \int_\Omega |2m \cdot \nabla u|^2 + N^2u^2 + 4Nu(m \cdot \nabla u)\ dx$$
$$= \int_\Omega |2m \cdot \nabla u|^2 + N^2u^2 + 2Nm \cdot \nabla(u^2)\ dx$$
$$= \int_\Omega |2m \cdot \nabla u|^2 + N^2u^2 - 2N^2u^2\ dx$$
$$+ 2N\int_\Gamma (m \cdot \nu)u^2\ d\Gamma.$$

We conclude by observing that

$$N^2 - 2N\mathrm{div}\, m = -N^2.$$

The following lemma is stronger than part (a) and the direct inequality in part (b) of Proposition 6.13.

Lemma 6.16.
 (a) The solutions of (6.49) satisfy the estimates

$$\int_0^T \int_\Gamma |\partial_\nu \Delta w|^2 + |\partial_\nu w'|^2 \, d\Gamma \, dt \le c \int_0^T \|f\|_{H_0^1(\Omega)}^2 \, dt$$

for every $T > 0$, with a constant c depending only on T.
 (b) The solutions of (6.50) satisfy the estimates

$$\int_0^T \int_\Gamma |\partial_\nu \Delta w|^2 + |\partial_\nu w'|^2 \, d\Gamma \, dt \le c \int_\Omega |\nabla \Delta w_0|^2 + |\nabla w_1|^2 \, dx$$

for every $T > 0$, with a constant c depending only on T.
 (c) The solutions of (6.50) satisfy the estimates

$$\int_\Omega |\nabla \Delta w_0|^2 + |\nabla w_1|^2 \, dx \le c \int_0^T \int_\Gamma |\partial_\nu \Delta w|^2 + |\partial_\nu w'|^2 \, d\Gamma \, dt$$

for every $T > R/\sqrt{\gamma_1}$, with a constant c depending only on T.
 Moreover, if we fix a positive integer k' and we consider only solutions whose initial data w_0, w_1 are orthogonal to e_k for every $k < k'$, then the above estimates hold for every $T > R/\sqrt{\gamma_{k'}}$, with a constant c depending only on k' and T.

Proof. By considering separately the real and imaginary parts of the solutions, it suffices to consider real-valued solutions. As a consequence of assumption (6.48), the left-hand side of (6.52) is minorized by a positive constant multiple of

$$\int_0^T \int_\Gamma (\partial_\nu \Delta w)^2 + |\partial_\nu w'|^2 \, d\Gamma \, dt.$$

Furthermore, putting

$$E := \frac{1}{2} \int_\Omega |\nabla \Delta w|^2 + |\nabla w'|^2 \, dx,$$

applying the preceding lemma, and using the variational characterization of the first Dirichlet eigenvalue γ_1 of $-\Delta$ in Ω, we have

$$\left| \int_{\Omega} w' M w \, dx \right| \leq \|w'\|_{L^2(\Omega)} \|Mw\|_{L^2(\Omega)}$$

$$\leq 2R\|w'\|_{L^2(\Omega)} \|\Delta w\|_{L^2(\Omega)}$$

$$\leq \frac{2R}{\sqrt{\gamma_1}} \|w'\|_{L^2(\Omega)} \|\nabla \Delta w\|_{L^2(\Omega)}$$

$$\leq \frac{2R}{\sqrt{\gamma_1}} E.$$

Therefore, using the preceding lemma again, the right-hand side of the identity (6.52) is majorized by

$$\frac{2R}{\sqrt{\gamma_1}} E(0) + \frac{2R}{\sqrt{\gamma_1}} E(T) + 4 \int_0^T E(t) \, dt + \frac{2R}{\sqrt{\gamma_1}} \int_0^T \|f(t)\|_{H_0^1(\Omega)} \sqrt{2E(t)} \, dt$$

and hence by

$$\left(\frac{4R}{\sqrt{\gamma_1}} + 4T \right) \|E\|_{L^\infty(0,T)} + \frac{2R}{\sqrt{\gamma_1}} \|f\|_{L^1(0,T;H_0^1(\Omega))} \|2E\|_{L^\infty(0,T)}^{1/2}.$$

If $w_0 = w_1 = 0$, then using the variation of constants formula for the associated abstract first-order problem with $F = (0, f)$ (see the remark following Theorem 2.1 on p. 10 and take into account that we are in the skew-adjoint case), we obtain the estimate

$$\|2E\|_{L^\infty(0,T)}^{1/2} \leq \|f\|_{L^1(0,T;H_0^1(\Omega))}.$$

Therefore, now the right-hand side of the identity (6.52) is majorized by

$$\left(4T + \frac{6R}{\sqrt{\gamma_1}} \right) \|f\|_{L^1(0,T;H_0^1(\Omega))}^2,$$

as required for part (a).

If $f = 0$, then the energy E is conserved because we are in the skew-adjoint case, so that the last expression reduces to

$$\left(4T + \frac{4R}{\sqrt{\gamma_1}} \right) E(0) = \left(2T + \frac{2R}{\sqrt{\gamma_1}} \right) \int_{\Omega} |\nabla \Delta w_0|^2 + |\nabla w_1|^2 \, dx.$$

This completes the proof of (b).

For the proof of (c) we observe that the left-hand side of the identity (6.52) is majorized by R times the left-hand side of the required inequality, while its right-hand side is minorized by

$$\left(4T - \frac{4R}{\sqrt{\gamma_1}} \right) E(0).$$

We conclude by remarking that the factor of $E(0)$ is strictly positive if $T > R/\sqrt{\gamma_1}$.

Moreover, if w_0, w_1 are orthogonal to e_k for every $k < k'$, then we can replace γ_1 by $\gamma_{k'}$ by the variational characterization of the eigenvalues. Therefore, the above estimates hold under the weaker condition $T > R/\sqrt{\gamma_{k'}}$.

In order to complete our proof of part (b) of Proposition 6.13, we shall apply the abstract Theorem 6.2 (p. 93). Furthermore, we shall also need the well-known fact that if $\varphi : \mathbb{R} \to \mathbb{R}$ is an even function of class C^∞ with compact support, then its Fourier transform $\Phi : \mathbb{R} \to \mathbb{R}$ defined by

$$\Phi(x) := \int_{-\infty}^{\infty} \varphi(t) e^{ixt} \, dt$$

is also an even function of class C^∞; moreover, it also belongs to the Schwartz space $\mathcal{S}$, so that

$$|x|^\alpha \Phi(x) \to 0$$

as $x \to \pm\infty$, for every fixed positive number α.

Finally, we also recall, e.g., from Agmon [1], that

$$\sum_{n=1}^{\infty} |\gamma_n|^{-N} < \infty. \tag{6.56}$$

Proof of part (b) of Proposition 6.10. Let us rewrite the problem (6.50) in the abstract form

$$U' = \mathcal{A}U, \quad U(0) = U_0, \tag{6.57}$$

as in the proof Proposition 3.5 (p. 48). Since now $\omega_k = \gamma_k$ because $a = 0$, we have

$$U(t) = \sum_{k=1}^{\infty} (U_k e^{i\gamma_k t} E_k + U_{-k} e^{-i\gamma_k t} E_{-k}), \quad U_{\pm k} \in \mathbb{C},$$

$$E_{\pm k} = \frac{1}{\sqrt{2}\gamma_k^{3/2}} (e_k, \pm i\gamma_k e_k),$$

$$\|(w, z)\| = \left(\int_\Omega |\nabla \Delta w|^2 + |\nabla z|^2 \, dx \right)^{1/2},$$

$$p(w, z) = \left(\int_\Gamma |\nabla \Delta w|^2 \, d\Gamma \right)^{1/2}.$$

Note that

$$\|U_0\|^2 = \sum_{k=1}^{\infty} |U_k|^2 + |U_{-k}|^2$$

and that

$$p(E_{\pm k}) \leq c_1 \sqrt{\gamma_k} \quad \text{for all} \quad k$$

with a constant c_1 by a classical trace theorem.

We are going to apply Theorem 6.2 with $m = 1$ and $p_1 = p$. It follows from part (b) of the preceding lemma that

$$\int_0^T p(U(t))^2 \, dt \leq c\|U_0\|^2$$

for all solutions of (6.57), for every $T > 0$. In particular, hypothesis (ii) of the theorem and the direct inequality in hypothesis (i) are satisfied for any choice of the set A. We complete our proof by showing that for any fixed $T > 0$, there exists a positive integer k' such that the inverse inequality in hypothesis (i) is satisfied with $A = \{1, \ldots, k' - 1\}$.

Fix an interval I whose closure is contained in $J := (0, T)$, and then fix an even function of class C^∞ satisfying the following conditions:

$$0 \leq \varphi \leq 1 \quad \text{in} \quad \mathbb{R},$$
$$\varphi = 1 \quad \text{on} \quad I,$$
$$\varphi = 0 \quad \text{outside} \quad J.$$

According to our above remark, its Fourier transform satisfies the inequality

$$|\Phi(x)| \leq c_2 |x|^{-N-1} \quad \text{for all} \quad x \neq 0$$

with a suitable constant c_2.

Fix a positive integer k', to be chosen later, and consider a solution $U(t)$ whose initial data are orthogonal to Z_k for all $k < k'$. Then we may write $U = U_+ + U_-$ with

$$U_+(t) := \sum_{k=k'}^\infty U_k e^{i\gamma_k t} E_k \quad \text{and} \quad U_+(t) := \sum_{k=k'}^\infty U_{-k} e^{-i\gamma_k t} E_{-k}.$$

We have, following Lebeau [92],

$$\int_J p(U(t))^2 \, dt \geq \int_{-\infty}^\infty \varphi(t) p(U(t))^2 \, dt$$

$$= \int_{-\infty}^\infty \varphi(t) p(U_+(t))^2 \, dt + \int_{-\infty}^\infty \varphi(t) p(U_-(t))^2 \, dt$$

$$+ 2 \int_{-\infty}^\infty \varphi(t) p(U_+(t), U_-(t)) \, dt$$

$$\geq \int_I p(U_+(t))^2 \, dt + \int_I p(U_-(t))^2 \, dt$$

$$+ 2 \int_{-\infty}^\infty \varphi(t) p(U_+(t), U_-(t)) \, dt.$$

Now observe that $U_+(t)$ and $U_-(t)$ are also solutions of (6.50) (with other initial data). Since

$$U'_+ = i\Delta U_+ \quad \text{and} \quad U'_- = -i\Delta U_-,$$

applying part (c) of the last lemma with a sufficiently large k' we have

$$\int_I p(U_+(t))^2 \, dt + \int_I p(U_-(t))^2 \, dt \geq c_3 \left(\|U_+(0)\|^2 + \|U_-(0)\|^2 \right) = c_3 \|U_0\|^2$$

with a positive constant c_3 that does not depend on the choice of k'. Therefore, we deduce from the above inequality the following estimate:

$$\int_J p(U(t))^2 \, dt \geq c_3 \|U_0\|^2 + 2 \int_{-\infty}^{\infty} \varphi(t) p(U_+(t), U_-(t)) \, dt.$$

Let us majorize the last integral. In the following sums, k and n run over the positive integers $k \geq k'$. We have

$$\left| \int_{-\infty}^{\infty} \varphi(t) p(U_+(t), U_-(t)) \, dt \right|$$

$$= \left| \sum_k \sum_n \Phi(\gamma_k + \gamma_n) p(U_k E_k, U_{-n} E_{-n}) \right|$$

$$\leq c_2 \sum_k \sum_n |\gamma_k + \gamma_n|^{-N-1} \left(p(U_k E_k)^2 + p(U_{-n} E_{-n})^2 \right)$$

$$\leq c_1^2 c_2 \sum_k \sum_n |\gamma_k + \gamma_n|^{-N-1} \left(\gamma_k |U_k|^2 + \gamma_n |U_{-n}|^2 \right)$$

$$\leq c_1^2 c_2 \sum_k \sum_n |\gamma_k + \gamma_n|^{-N} \left(|U_k|^2 + |U_{-n}|^2 \right)$$

$$= c_1^2 c_2 \sum_k \left(\sum_n |\gamma_k + \gamma_n|^{-N} \right) \left(|U_k|^2 + |U_{-k}|^2 \right).$$

Putting

$$c_4(k') := \sum_{n=k'}^{\infty} |\gamma_n|^{-N},$$

we conclude that

$$\left| \int_{-\infty}^{\infty} \varphi(t) p(U_+(t), U_-(t)) \, dt \right| \leq c_1^2 c_2 c_4(k') \|U_0\|^2,$$

and therefore

$$\int_J p(U(t))^2 \, dt \geq \left(c_3 - c_1^2 c_2 c_4(k') \right) \|U_0\|^2.$$

As a consequence of the relation (6.56) we have $c_4(k') \to 0$ as $k' \to \infty$. Therefore, the factor of $\|U_0\|^2$ is strictly positive if k' is chosen to be sufficiently large at the beginning of the proof.

7

Problems on Spherical Domains

This chapter is devoted to the study of some higher-dimensional problems that can be solved by the method developed in Chapter 4. For the convenience of the reader we recall briefly the description of the eigenfunctions of the Laplacian operator in balls, and we also give a very short and elementary introduction to the Bessel functions, by establishing all properties we need in this book. We also establish some new results concerning the zeros of Bessel-type functions, and we present new, simpler proofs of some classical results.

7.1 Observability of the Wave Equation in a Ball

Let us consider the wave equation with Neumann boundary condition as in Section 3.3 (p. 42):

$$\begin{cases} u'' - \Delta u + au = 0 & \text{in} \quad \mathbb{R} \times \Omega, \\ \partial_\nu u = 0 & \text{on} \quad \mathbb{R} \times \Gamma, \\ u(0) = u_0 \quad \text{and} \quad u'(0) = u_1 & \text{in} \quad \Omega. \end{cases} \tag{7.1}$$

We recall from Proposition 3.3 (p. 43) that this problem is well-posed for $u_0 \in H^1(\Omega)$ and $u_1 \in L^2(\Omega)$. As usual, we consider only solutions with initial data belonging to the dense subspace $\mathcal{Z} := Z \times Z$, where Z denotes the linear hull of the eigenfunctions of $-\Delta$ with homogeneous Neumann boundary condition in Ω.

Defining the energy by the formula

$$E(t) := \frac{1}{2} \int_\Omega |u(t,x)|^2 + |\nabla u(t,x)|^2 + |u'(t,x)|^2 \; dx, \quad t \in \mathbb{R},$$

it is natural to conjecture that

$$\int_I \int_\Gamma |u|^2 + |u'|^2 \; d\Gamma \; dt \asymp E(0).$$

for every sufficiently large interval. Indeed, this has been established for the one-dimensional case in Proposition 4.11 (p. 76). It turns out, however, that this relation is inexact in several dimensions:

Proposition 7.1. *Let Ω be an open ball of radius R in $\mathbb{R}^N$ with $N \geq 2$.*

(a) The inverse *inequality*

$$E(0) \leq c \int_I \int_\Gamma |u|^2 + |u'|^2 \; d\Gamma \; dt \tag{7.2}$$

holds for all solutions of (7.1) with $(u_0, u_1) \in \mathcal{Z}$ if $|I| > 2R$, with a constant c depending on I.

(b) The direct *inequality*

$$\int_I \int_\Gamma |u|^2 + |u'|^2 \; d\Gamma \; dt \leq cE(0) \tag{7.3}$$

can fail for any interval $|I|$.

Remarks.

- For $a = 0$ this is essentially due to Graham and Russell [44]. We shall give a simplified proof.
- The analysis of our proof shows that the inverse inequality (7.2) can also fail if $|I| < 2R$. A deeper theorem of Joó [59], [60] states that for $a = 0$ in dimensions $N \geq 2$ the inverse inequality (7.2) can fail in the limiting case $|I| = 2R$, too. This contrasts with the one-dimensional case: the proofs of both Proposition 1.1 and Proposition 4.14 (pp. 2, 81) can be easily adapted to show that the estimates of Proposition 4.11 (p. 76) hold for all intervals of length ℓ if $a = 0$.
- Our proof will also show that if I_1 and I_2 are two intervals of length $> 2R$, then the corresponding observations are equivalent:

$$\int_{I_1} \int_\Gamma |u|^2 + |u'|^2 \; d\Gamma \; dt \asymp \int_{I_2} \int_\Gamma |u|^2 + |u'|^2 \; d\Gamma \; dt.$$

  This property enables us to apply the abstract stabilization Theorem 2.14 (p. 24); see [70] for the details.
- Arguments of geometric optics indicate[1] that the just-mentioned equivalence can fail for general domains Ω. Since the methods of multipliers, Carleman estimates, and microlocal analysis are not very sensitive to deformations of the domain, we do not see other means to establish the equivalence property for balls.
- The proof of Proposition 7.1 can easily be adapted to the case of Dirichlet boundary condition. In this way we can get a new proof of part (b) of Proposition 6.9 (p. 110), but only in the special case of balls. Here the above-mentioned other methods are more efficient.

---

[1]C. Bardos, private communication.

The proof of Proposition 7.1 is based on the explicit representation of the eigenfunctions of the Laplacian operator in balls. Therefore, we first recall in the next section the description of these eigenfunctions by using Bessel functions and spherical harmonics. Then in Section 7.3 we present some classical and recent results concerning the location of the zeros of Bessel-type functions. Using this, Proposition 7.1 will be proved in Section 7.4.

7.2 The Eigenfunctions of the Laplacian Operator in Balls

This review section is devoted to a description of the eigenfunctions of the Laplacian operator with Dirichlet or Neumann boundary condition in the unit ball Ω_1 of $\mathbb{R}^N$ ($N \geq 2$) with boundary Γ_1. We omit the proofs, and we refer to [19], [27], [127], [129], and [135] for details.

In order to formulate our results, let us introduce the Bessel functions of any real order m by the formula

$$J_m(x) = \sum_{j=0}^{\infty} \frac{(-1)^j}{j! \Gamma(m+j+1)} \left(\frac{x}{2}\right)^{m+2j}, \quad x \geq 0. \tag{7.4}$$

Examples.

- One can readily verify that

$$J_0'(x) = -J_1(x) \quad \text{and} \quad \frac{d}{dx}(x J_1(x)) = x J_0(x).$$

- If m is half an odd integer, then $J_m(x)$ can be expressed in a finite form. For example,

$$J_{-1/2}(x) = \sqrt{\frac{2}{\pi x}} \cos x \quad \text{and} \quad J_{1/2}(x) = \sqrt{\frac{2}{\pi x}} \sin x.$$

The asymptotic behavior of $J_m(x)$ is similar for all indices:

Lemma 7.2. *For every real number m we have*

$$J_m(x) = \sqrt{\frac{2}{\pi x}} \cos\left(x - \frac{m\pi}{2} - \frac{\pi}{4}\right) + O(x^{-3/2})$$

as $x \to \infty$. Furthermore, for any given pair $(\alpha, \beta) \neq (0,0)$ of real numbers, the positive zeros of $\alpha J_m(x) + \beta x J_m'(x)$ form an infinite sequence tending to ∞. (See Figures 7.1–7.8.)

Let us also recall that the *spherical harmonics* of order m ($m = 0, 1, \dots$) are the restrictions to the unit sphere Γ_1 of the homogeneous polynomials of order m.

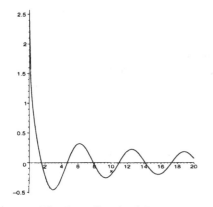

Fig. 7.1. Graph of $J_{-1/2}$

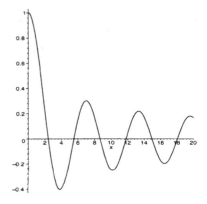

Fig. 7.2. Graph of J_0

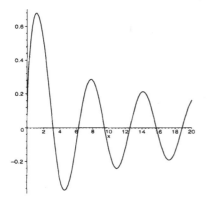

Fig. 7.3. Graph of $J_{1/2}$

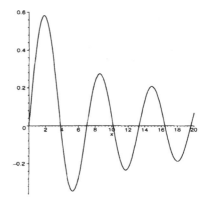

Fig. 7.4. Graph of J_1

Lemma 7.3. *The spherical harmonics of order m form a finite-dimensional subspace $\mathcal{S}_m$ in $L^2(\Gamma_1)$. These subspaces are mutually orthogonal, and their linear hull is dense in $L^2(\Gamma_1)$.*

Examples.

- If $N = 2$, then $\mathcal{S}_m$ is spanned by the functions $e^{\pm im\theta}$ with the usual parameterization of the unit circle Γ_1. Hence $\mathcal{S}_0$ is one-dimensional, while the other subspaces are two-dimensional.
- If $N = 3$, then dim $\mathcal{S}_m = 2m + 1$ for all m, and its elements can be conveniently expressed using the Legendre polynomials.
- In the general case we have dim $\mathcal{S}_0 = 1$, dim $\mathcal{S}_1 = N$, and

$$\dim \mathcal{S}_m = \binom{N + m - 1}{m} - \binom{N + m - 3}{m - 2}$$

for $m \geq 2$. This follows from the fact that every spherical harmonic function of order m is also the restriction to the unit sphere Γ of a unique

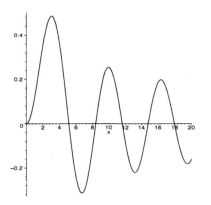

Fig. 7.5. Graph of J_2

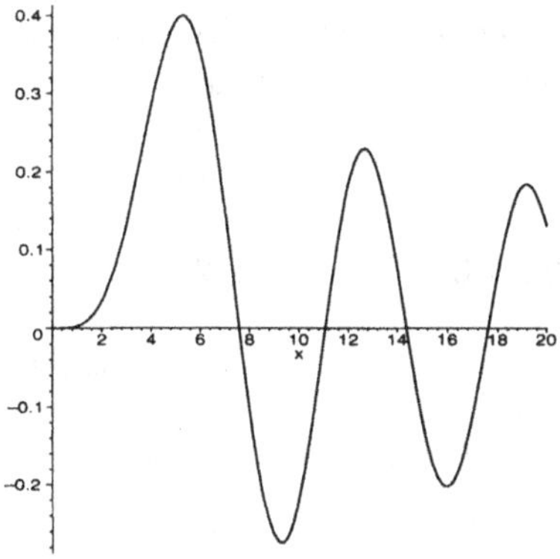

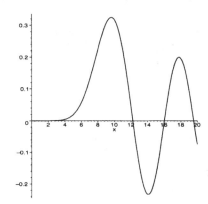

Fig. 7.6. Graph of J_4

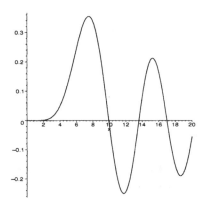

Fig. 7.7. Graph of J_6

Fig. 7.8. Graph of J_8

harmonic homogeneous polynomial of order m. The elements of $\mathcal{S}_m$ can be expressed by the ultraspherical polynomials.

Now we are ready to describe the eigenfunctions of the Laplacian operator. We use hyperspherical (polar) coordinates (r, θ) with $0 \le r \le 1$ and $\theta \in \Gamma_1$.

Proposition 7.4. *Consider the unit ball Ω_1 of $\mathbb{R}^N$ ($N \ge 2$) with boundary Γ_1.*

(a) The eigenfunctions of $-\Delta$ in Ω_1 with Dirichlet boundary condition are exactly the functions

$$(r, \theta) \mapsto r^{1-\frac{N}{2}} J_{m-1+\frac{N}{2}}(c_{m,k} r) H_m(\theta),$$

where $m = 0, 1, \ldots$, $k = 1, 2, \ldots$, $H_m \in \mathcal{S}_m$, and for each m we denote by

$$0 < c_{m,1} < c_{m,2} < \cdots$$

the sequence of positive zeros of the Bessel function $J_{m-1+\frac{N}{2}}(x)$. The corresponding eigenvalue is equal to $c_{m,k}^2$.

(b) The nonconstant eigenfunctions of $-\Delta$ in Ω_1 with Neumann boundary condition are exactly the functions

$$(r, \theta) \mapsto r^{1 - \frac{N}{2}} J_{m-1+\frac{N}{2}}(c_{m,k} r) H_m(\theta),$$

where $m = 0, 1, \ldots,$ $k = 1, 2, \ldots,$ $H_m \in \mathcal{S}_m,$ and for each m we denote by

$$0 < c_{m,1} < c_{m,2} < \cdots$$

the sequence of positive zeros of the function

$$\left(1 - \frac{N}{2}\right) J_{m-1+\frac{N}{2}}(x) + x J'_{m-1+\frac{N}{2}}(x).$$

The corresponding eigenvalue is equal to $c_{m,k}^2$.

Example. In the three-dimensional case, the smallest Dirichlet eigenvalue is equal to $c_{0,1}^2 = \pi^2$, and the corresponding eigenspace is spanned by the positive radial function

$$(r, \theta) \mapsto \frac{\sin \pi r}{r}.$$

7.3 Zeros of Bessel-type Functions

The rest of this section is devoted to the study of the zeros $c_{m,k}$ figuring in Proposition 7.4. Let us first recall some elementary facts.

Lemma 7.5. *Let m be a nonnegative real number.*

(a) The following equality holds for every positive real number c:

$$2c^2 \int_0^1 r |J_m(cr)|^2 \, dr = c^2 |J'_m(c)|^2 + (c^2 - m^2)|J_m(c)|^2. \tag{7.5}$$

(b) The functions $J_m(x)$ and $J'_m(x)$ are positive in $(0, m]$, so their first positive roots are bigger than m.

(c) If $\alpha \geq 0$, then the first positive root of $\alpha J_m(x) + x J'_m(x)$ is bigger than m.

(d) If $\alpha < 0$ and $m > |\alpha|$, then the first positive root of $\alpha J_m(x) + x J'_m(x)$ is bigger than $\sqrt{m^2 - \alpha^2}$.

Proof.

(a) Using the power series expansion (7.4), we see that $y := J_m(x)$ satisfies the differential equation

$$x^2 y'' + x y' + (x^2 - m^2)y = 0 \quad \text{in} \quad (0, \infty). \tag{7.6}$$

Multiplying this equation by $2y'$ and integrating over $(0, c)$, we obtain the equality

$$\int_0^c 2xy^2 \, dx = \left[x^2 (y')^2 + (x^2 - m^2) y^2 \right]_0^c = c^2 y'(c)^2 + (c^2 - m^2) y(c)^2 + m^2 y(0)^2.$$

Since $y(0) = 0$ for $m > 0$ from the power series expansion (7.4), the last term always vanishes. The equality (7.5) follows by the change of variables $x = cr$ in the integral.

(b) One can see from the power series expansion (7.4) that $J_m(x)$ and $J_m'(x)$ are positive for small positive values of x. Therefore, we infer from (7.5) that

$$|J_m'(c)|^2 \geq 2 \int_0^1 r |J_m(cr)|^2 \, dr > 0$$

for all $0 < c \leq m$. Hence $J_m'(x)$ is positive in $(0, m]$. Consequently, $J_m(x)$ is increasing and thus also positive in $(0, m]$.

(c) If $\alpha \geq 0$, then $\alpha J_m(x) + x J_m'(x)$ is positive in $(0, m]$ by property (b).

(d) If $\alpha J_m(c) + c J_m'(c) = 0$ for some $0 < c \leq m$, then we obtain from (7.5) the following equality:

$$2c^2 \int_0^1 r |J_m(cr)|^2 \, dr = (\alpha^2 + c^2 - m^2) |J_m(c)|^2.$$

Since $J_m > 0$ in $(0, m]$ by (b), we must have $\alpha^2 + c^2 - m^2 > 0$, i.e., $c > \sqrt{m^2 - \alpha^2}$.

Next we recall the classical Sturm oscillation theorem:

Theorem 7.6. *Let $f, g : [a, b] \rightarrow \mathbb{R}$ be two continuous functions satisfying*

$$f < g \quad \text{on} \quad (a, b), \tag{7.7}$$

and let $u, v \in H^2(a, b)$ be two functions satisfying the differential equations

$$u'' + fu = 0 \quad \text{and} \quad v'' + gv = 0 \quad \text{in} \quad (a, b).$$

Assume that

$$u(a) = u(b) = 0 \tag{7.8}$$

and that u does not vanish in (a, b). Then v has at least one zero in (a, b).

Sturm's theorem states that the greater the value of f, the more rapidly the solutions oscillate.

Proof. Assume to the contrary that v does not vanish in (a, b). Then, changing u to $-u$ and/or v to $-v$ if necessary, we may assume that

$$u > 0 \text{ and } v > 0 \quad \text{in} \quad (a, b). \tag{7.9}$$

Then we also have

$$v(a) \geq 0, \quad v(b) \geq 0, \quad u'(a) \geq 0, \quad \text{and} \quad u'(b) \leq 0. \tag{7.10}$$

Multiplying the differential equations by v and u, respectively, and integrating their difference, we obtain

$$[uv' - u'v]_a^b + \int_a^b (g - f)uv \, dx = 0;$$

using (7.8) this simplifies to

$$u'(a)v(a) - u'(b)v(b) + \int_a^b (g - f)uv \, dx = 0.$$

But this is impossible, because $u'(a)v(a) - u'(b)v(b) \geq 0$ by (7.10), while the integral is (strictly) positive by (7.7) and (7.9).

Corollary 7.7. *Let $h : [d, \infty) \to \mathbb{R}$ be a continuous function, having a positive limit L at infinity. Let y be a nonzero solution of the differential equation*

$$y'' + hy = 0 \quad \text{in} \quad (d, \infty).$$

Then y has an infinite sequence of zeros $c_1 < c_2 < \cdots$, all simple, tending to infinity and satisfying the relation

$$c_{n+1} - c_n \to \pi/\sqrt{L}.$$

Proof. If y had a multiple root c, then by the uniqueness of the solution of the initial value problem

$$y'' + hy = 0 \quad \text{in} \quad (d, \infty), \quad y(c) = y'(c) = 0,$$

we could conclude that y is identically zero, contradicting our assumption.

Since h is bounded by assumption, there exists a positive number β such that $h < \beta^2$ on $[d, \infty)$. We claim that the distance between any two zeros of y is at least equal to π/β. Indeed, if $y(c) = 0$, then we apply Sturm's theorem with the following choice:

$$(a, b) := (c, c + \pi/\beta);$$
$$f(x) := h(x) \quad \text{and} \quad u(x) := y(x);$$
$$g(x) := \beta^2 \quad \text{and} \quad v(x) := \sin \beta(x - c).$$

Since v does not vanish in (a, b), u cannot vanish in it either.

Next fix a sufficiently large number $a \geq d$ such that $h > 0$ in $[a, \infty)$, and choose a positive number α such that $\alpha^2 < h$ in $[a, \infty)$. Then y has at least one root in the interval $(a, a + \pi/\alpha)$. This can be shown by applying Sturm's theorem with the following choice:

$$(a, b) := (a, a + \pi/\alpha);$$
$$f(x) := \alpha^2 \quad \text{and} \quad u(x) := \sin \alpha(x - a);$$
$$g(x) := h(x) \quad \text{and} \quad v(x) := y(x).$$

Thus y has infinitely many zeros, and they can be arranged into an increasing sequence $c_1 < c_2 < \cdots$, tending to infinity.

Now choose two positive numbers α and β, arbitrarily close to $\sqrt{L}$ and satisfying the ineqalities $\alpha < \sqrt{L} < \beta$. Choose a sufficiently large $a \geq d$ such that $\alpha^2 < h < \beta^2$ in $[a, \infty)$; then the above arguments also show that

$$\pi/\beta < c_{n+1} - c_n < \pi/\alpha$$

for all sufficiently large indices n such that $c_n \geq a$. This completes the proof.

As a first illustration, let us recall Porter's classical proof on the location of the zeros of the Bessel functions; our proof of part (d) seems to be new:

Proposition 7.8.

(a) *For any given real number m, the positive zeros of $J_m(x)$ are simple, and they form an infinite increasing sequence $j_1 < j_2 < \cdots$, tending to infinity.*

(b) *The difference sequence $(j_{n+1} - j_n)$ converges to π.*

(c) *The sequence $(j_{n+1} - j_n)$ is strictly decreasing if $|m| > 1/2$, strictly increasing if $|m| < 1/2$, and constant if $m = \pm 1/2$.*

(d) *The first positive zero j_1 of $J_m(x)$ satisfies the inequalities*

$$m + m^{1/3} < j_1 < m + 4m^{1/3}$$

if m is sufficiently large.

Remark. The proof of (d) will yield the stronger result

$$m + \beta m^{1/3} < j_1 < m + \gamma m^{1/3}$$

with any given $\beta < (\pi^2/2)^{1/3} \approx 1.7$ and $\gamma > 3\pi^{2/3}/2 \approx 3.2$ if m is sufficiently large. It was proved by deeper tools that

$$j_1 = m + \delta m^{1/3} + O(1), \quad m \to \infty,$$

with a constant $\delta \approx 1.855757$; see, e.g., Watson [135], p. 516.

Proof. Since the function $z(x) := J_m(x)$ satisfies

$$x^2 z'' + xz' + (x^2 - m^2)z = 0 \quad \text{in} \quad (0, \infty),$$

(see (7.6), p. 132), the function $y(x) := \sqrt{x} J_m(x)$ satisfies

$$y'' + hy = 0 \quad \text{in} \quad (0, \infty) \quad \text{with} \quad h(x) := 1 - \frac{m^2 - \frac{1}{4}}{x^2}. \tag{7.11}$$

Note that $y(x)$ and $J_m(x)$ have the same positive roots. The cases $m = \pm 1/2$ of the proposition hence follow at once because then $h(x) = 1$, so that $y(x)$ is a suitable constant multiple of the trigonometric function $\sin(x + \gamma)$ for some γ.

In the remaining cases fix a small $d > 0$ so that all positive roots of $J_m(x)$ lie in (d, ∞). (This is possible because $J_m(x) > 0$ for small positive values of x.) Then the function h satisfies the hypotheses of the preceding corollary, and properties (a) and (b) follow.

Property (c) will be proved by applying Sturm's theorem. If $|m| > 1/2$, then h is strictly increasing in $[d, \infty)$. If $j_n < j_{n+1} < j_{n+2}$ are three consecutive roots of y, then we apply Sturm's theorem with the following choice:

$$(a, b) := (j_n, j_{n+1});$$
$$f(x) := h(x) \quad \text{and} \quad u(x) := y(x);$$
$$g(x) := h(x + j_{n+1} - j_n) \quad \text{and} \quad v(x) := y(x + j_{n+1} - j_n).$$

We conclude that v has a root in (j_n, j_{n+1}); i.e., y has a root in $(j_{n+1}, 2j_{n+1} - j_n)$. This implies that $j_{n+2} < 2j_{n+1} - j_n$, which is equivalent to the inequality

$$j_{n+2} - j_{n+1} < j_{n+1} - j_n.$$

If $|m| < 1/2$, then h is strictly decreasing. If $j_n < j_{n+1} < j_{n+2}$ are three consecutive roots of y, then we apply Sturm's theorem in the following way:

$$(a, b) := (j_{n+1}, j_{n+2});$$
$$f(x) := h(x) \quad \text{and} \quad u(x) := y(x);$$
$$g(x) := h(x - j_{n+1} + j_n) \quad \text{and} \quad v(x) := y(x - j_{n+1} + j_n).$$

We conclude that v has a root in (j_{n+1}, j_{n+2}); i.e., y has a root in $(j_n, j_{n+2} - j_{n+1} + j_n)$. This implies that $j_{n+1} < j_{n+2} - j_{n+1} + j_n$, which is equivalent to the inequality

$$j_{n+1} - j_n < j_{n+2} - j_{n+1}.$$

Turning to the proof of (d), first we show that for every fixed $\alpha > \pi^{2/3}/2$, $J_m(x)$ has at least one root in the interval

$$(a, b) := \left(m + \alpha m^{1/3}, m + 3\alpha m^{1/3}\right)$$

if m is large enough. We are going to apply Sturm's theorem with the following choice:

$$f(x) := \frac{\pi^2}{4\alpha^2 m^{2/3}} \quad \text{and} \quad u(x) := \sin \frac{\pi}{2\alpha m^{1/3}} \left(x - m - \alpha m^{1/3}\right),$$
$$g(x) := h(x) \quad \text{and} \quad v(x) := y(x).$$

We have only to verify the condition $g > f$ in (a, b).

Since g is increasing for $m > 1/2$ and since f is a positive constant, it suffices to show that $g(a)/f(a) > 1$. A straightforward computation shows that

$$\frac{g(a)}{f(a)} = \frac{4\alpha^2 m^{2/3}}{\pi^2} \left(1 - \frac{m^2 - \frac{1}{4}}{\left(m + \alpha m^{1/3}\right)^2}\right)$$

$$= \frac{4\alpha^2 m^{2/3}}{\pi^2} \frac{\left(m + \alpha m^{1/3}\right)^2 - \left(m^2 - \frac{1}{4}\right)}{\left(m + \alpha m^{1/3}\right)^2}$$

$$= \frac{4\alpha^2 m^{2/3}}{\pi^2} \frac{2\alpha m^{4/3} + \alpha^2 m^{2/3} + \frac{1}{4}}{m^2 + 4\alpha m^{4/3} + \alpha^2 m^{2/3}}$$

$$\to \frac{8\alpha^3}{\pi^2}$$

as $m \to \infty$. We conclude by remarking that the last limit is bigger than one by the choice of α.

Next we prove that for any fixed $\alpha < (\pi^2/2)^{1/3}$, $J_m(x)$ has no root in the interval

$$(a, b) := \left(m, m + \alpha m^{1/3}\right)$$

if m is sufficiently large. For this we apply Sturm's theorem with

$$f(x) := h(x) \quad \text{and} \quad u(x) := y(x),$$

$$g(x) := \frac{\pi^2}{\alpha^2 m^{2/3}} \quad \text{and} \quad v(x) := \sin \frac{\pi}{\alpha m^{1/3}}(x - m).$$

The condition $g > f$ in (a, b) is equivalent to $f(b)/g(b) < 1$. By repeating the above computation, we have

$$\frac{f(b)}{g(b)} = \frac{\alpha^2 m^{2/3}}{\pi^2} \left(1 - \frac{m^2 - \frac{1}{4}}{\left(m + \alpha m^{1/3}\right)^2}\right) \to \frac{2\alpha^3}{\pi^2}$$

as $m \to \infty$. We conclude by remarking that the last limit is smaller than one by our assumption on α.

Now we study, following [65], the zeros of the functions $\alpha J_m(x) + x J'_m(x)$. It will be used in the next section.

Proposition 7.9. *Fix a real number α.*

(a) For each $m \geq 0$, the positive zeros of $\alpha J_m(x) + x J'_m(x)$ are simple, and they form an infinite increasing sequence $c_1 < c_2 < \cdots$, tending to infinity.

(b) The difference sequence $(c_{n+1} - c_n)$ converges to π.

(c) If m is sufficiently large, then the sequence $(c_{n+1} - c_n)$ is strictly decreasing.

(d) If $\alpha \leq 0$, then $\sqrt{m^2 - \alpha^2} < c_1 < m + 4m^{1/3}$ for all sufficiently large m, so that $c_1/m \to 1$ as $m \to \infty$.

Remark. We refer to [65] for more general results, and to a recent work of Oudet [110], where the zeros of higher-order derivatives of Bessel functions were also studied.

First we construct a function U having the same zeros as $\alpha J_m(x) + x J_m'(x)$, for which we can apply Sturm's theorem. We need the following lemma:

Lemma 7.10. *Let I be an open interval, $f, g : I \to \mathbb{R}$ two functions of class C^3, and let u be a solution of the differential equation*

$$(gu)'' + fgu = 0 \quad \text{in} \quad I. \tag{7.12}$$

Assume that g and $h := g'' + fg$ have no zeros in I, and let us introduce two new functions by setting

$$G := |g^3/h|^{1/2}$$

and

$$F := f + 3(g''/g) - 2(g'/g)(h'/h) - (G''/G).$$

Then G has no zeros in I, and

$$(Gu')'' + FGu' = 0 \quad \text{in} \quad I. \tag{7.13}$$

Proof. Differentiating equation (7.12), we obtain

$$gu''' + 3g'u'' + (2g'' + h)u' + h'u = 0 \quad \text{in} \quad I.$$

Combining this result with (7.12), we eliminate u to obtain

$$gu''' + (3g' - gh'/h)u'' + (2g'' + h - 2g'h'/h)u' = 0 \quad \text{in} \quad I.$$

Comparing this equation with (7.13), we obtain for F and G the following conditions:

$$2G'/G = 3(g'/g) - (h'/h),$$
$$F + (G''/G) = 2(g''/g) + (h/g) - 2(g'/g)(h'/h).$$

The first condition is obviously satisfied by the above choice of G, while the second is just the definition of F.

Proof of Proposition 7.9. As a consequence of parts (c) and (d) of Lemma 7.5 (p. 132) we may fix a positive number d such that $d^2 + \alpha^2 - m^2 > 0$ and such that all positive zeros of $y(x) := \alpha J_m(x) + x J_m'(x)$ lie in (d, ∞).

It follows from equation (7.11) that (7.12) is satisfied on the interval $I = (d, \infty)$ with the following choice:

$$f(x) := 1 - \frac{m^2 - \frac{1}{4}}{x^2},$$
$$g(x) := x^{\frac{1}{2} - \alpha},$$
$$u(x) := x^\alpha y(x).$$

Since

$$g(x) \quad \text{and} \quad h(x) := g''(x) + f(x)g(x) = x^{-\alpha - \frac{3}{2}}(x^2 + \alpha^2 - m^2)$$

do not vanish in this interval, we may apply the preceding lemma. It follows that

$$U(x) := (Gu')(x) = x^{\frac{3}{2} - \alpha}(x^2 + \alpha^2 - m^2)^{-1/2} u'$$

and

$$F(x) := 1 - \frac{m^2 - \frac{1}{4}}{x^2} + \frac{2\alpha - 1}{x^2 + \alpha^2 - m^2} + 3\frac{\alpha^2 - m^2}{(x^2 + \alpha^2 - m^2)^2}$$

satisfy

$$U'' + FU = 0 \quad \text{in} \quad (d, \infty).$$

Let us note that $U(x)$ has the same positive zeros as $\alpha J_m(x) + x J'_m(x)$. Since F satisfies the hypotheses of Corollary 7.7 (p. 134) with $L = 1$, parts (a) and (b) of the proposition follow. Part (c) will also be established if we show that (for any fixed α) the function F is strictly increasing for every sufficiently large value of m.

With the definition $z := x^2 + \alpha^2 - m^2$ for brevity, a straightforward computation leads to the identity

$$2x^3 z^3 F'(x) = Az^3 + Bz^2 + Cz + D$$

with

$$A = 4m^2 + 3 - 8\alpha,$$
$$B = 16(2 - \alpha)(m^2 - \alpha^2),$$
$$C = (52 - 8\alpha)(m^2 - \alpha^2)^2,$$
$$D = 24(m^2 - \alpha^2)^3.$$

We complete the proof of part (c) by showing that

$$Az^3 + Bz^2 + Cz + D > 0 \quad \text{for all} \quad z > 0$$

if m is sufficiently large.

Introducing the new variable $t := z/(m^2 - \alpha^2)$, we see that this is equivalent to

$$(4m^2 + 3 - 8\alpha)t^3 + (32 - 16\alpha)t^2 + (52 - 8\alpha)t + 24 > 0 \quad \text{for all} \quad t > 0 \quad (7.14)$$

if m is sufficiently large.

Let us first choose a sufficiently small $T > 0$ such that

$$(3 - 8\alpha)t^3 + (32 - 16\alpha)^2 + (52 - 8\alpha)t + 24 > 0 \quad \text{for all} \quad t \in [0, T]. \quad (7.15)$$

Then choose a sufficiently large M satisfying the following inequalities:

$$M > |\alpha|,$$
$$4M^2 + 3 - 8\alpha > 0,$$
$$(4M^2 + 3 - 8\alpha)T/3 + (32 - 16\alpha) > 0,$$
$$(4M^2 + 3 - 8\alpha)T^2/3 + (52 - 8\alpha) > 0,$$
$$(4M^2 + 3 - 8\alpha)T^3/3 + 24 > 0.$$

It follows from these inequalities that

$$(4m^2 + 3 - 8\alpha)t^3 + (32 - 16\alpha)t^2 + (52 - 8\alpha)t + 24 > 0 \qquad (7.16)$$

for all $t > T$ and $m \geq M$. Finally, (7.15) and (7.16) imply (7.14).

It remains to prove part (d) of the proposition. As a consequence of parts (d) of Lemma 7.5 and Proposition 7.8, it suffices to show that $y(x) := \alpha J_m(x) + x J_m'(x)$ changes sign between 0 and the first positive root j_1 of $J_m(x)$. First we observe that the first positive root j_1' of $J_m'(x)$ is smaller than j_1 if $m > 0$. Indeed, this follows from Rolle's theorem because $J_m(0) = 0$. Now we notice that $y(j_1') = \alpha J_m(j_1') < 0$ because $J_m(x) > 0$ between 0 and j_1. On the other hand, it follows easily from the power series representation (7.4), p. 129, of $J_m(x)$ that $y(x) > 0$ for small positive values of x if $m > |\alpha|$. Hence y changes sign between 0 and $j_1' < j_1$.

7.4 Proof of Proposition 7.1

We may assume by a scaling argument that Ω is the unit disk. Denoting by

$$c_{m,1} < c_{m,2} < \cdots$$

the positive roots of the function

$$\left(1 - \frac{N}{2}\right) J_{m-1+\frac{N}{2}}(x) + x J_{m-1+\frac{N}{2}}'(x), \qquad (7.17)$$

setting

$$c_{0,0} := 0,$$
$$\omega_{m,k} := \sqrt{c_{m,k}^2 + a}$$

and

$$R_{m,k}(r) := \begin{cases} 1 & \text{if } m = k = 0, \\ r^{1-\frac{N}{2}} J_{m-1+\frac{N}{2}}(c_{m,k}r) & \text{otherwise}, \end{cases}$$

it follows from Proposition 7.4 (p. 131) that the solutions of the problem

$$\begin{cases} u'' - \Delta u + au = 0 & \text{in } \mathbb{R} \times \Omega, \\ \partial_\nu u = 0 & \text{on } \mathbb{R} \times \Gamma, \\ u(0) = u_0 \text{ and } u'(0) = u_1 & \text{in } \Omega, \end{cases}$$

for $(u_0, u_1) \in \mathcal{Z}$ are given by finite sums of the form[2]

$$u(t, r, \theta) = \sum_m \sum_k R_{m,k}(r) \big(H_{m,k}^+(\theta) e^{i\omega_{m,k} t} + H_{m,k}^-(\theta) e^{-i\omega_{m,k} t} \big), \quad (7.18)$$

where $H_{m,k}^+$ and $H_{m,k}^-$ are suitable spherical harmonics of order m, depending on the initial data. Here and in the sequel, m runs over $0, 1, \ldots$, while k runs over $0, 1, \ldots$ if $m = 0$ and over $1, 2, \ldots$ if $m \geq 1$.

Applying Proposition 3.3 (p. 43), we obtain that

$$E(0) \asymp \sum_m \sum_k \alpha_{m,k} \int_\Gamma |H_{m,k}^+(\theta)|^2 + |H_{m,k}^-(\theta)|^2 \, d\Gamma \quad (7.19)$$

with

$$\alpha_{m,k} := (1 + c_{m,k}^2) \int_0^1 r^{N-1} R_{m,k}(r)^2 \, dr.$$

Since $c_{m,k}$ is a root of (7.17), using the identity (7.5) of Lemma 7.5 (p. 132), we obtain (except for the case $m = k = 0$) that

$$2c_{m,k}^2 \int_0^1 r^{N-1} R_{m,k}(r)^2 \, dr$$

$$= \left(\left(1 - \frac{N}{2}\right)^2 + c_{m,k}^2 - \left(m - 1 + \frac{N}{2}\right)^2 \right) |R_{m,k}(1)|^2$$

$$= \left(c_{m,k}^2 - m^2 + (N - 2)m \right) R_{m,k}(1)^2.$$

Hence

$$\alpha_{m,k} \asymp \left(1 + c_{m,k}^2 - m^2\right) R_{m,k}(1)^2. \quad (7.20)$$

Next we compute the boundary integral. Let us first assume that $\omega_{m,k} \neq 0$ for all (m, k). Using the orthogonality of spherical harmonics of different order, we have

$$\int_\Gamma |u(t, 1, \theta)|^2 \, d\Gamma$$

$$\asymp \sum_m \int_\Gamma \left| \sum_k R_{m,k}(1) \big(H_{m,k}^+(\theta) e^{i\omega_{m,k} t} + H_{m,k}^-(\theta) e^{-i\omega_{m,k} t} \big) \right|^2 \, d\Gamma.$$

Now fix an interval I of length > 2. As a consequence of Proposition 6.9 (p. 110) we may apply Ingham's Theorem 4.3 (p. 59) for all sufficiently large indices m, with the same constants in the estimates, so that

[2]If $\omega_{m,k} = 0$ for some (m, k), then $H_{m,k}^+(\theta) e^{i\omega_{m,k} t} + H_{m,k}^-(\theta) e^{-i\omega_{m,k} t}$ is replaced by $H_{m,k}^+(\theta) + H_{m,k}^-(\theta)t$.

$$\int_I \left| \sum_k R_{m,k}(1) \left(H^+_{m,k}(\theta) e^{i\omega_{m,k}t} + H^-_{m,k}(\theta) e^{-i\omega_{m,k}t} \right) \right|^2 dt$$

$$\asymp \sum_k R_{m,k}(1)^2 \left(\left| H^+_{m,k} \right|^2 + \left| H^-_{m,k} \right|^2 \right).$$

The same estimates are obtained for each of the remaining finitely many indices m by applying Theorem 4.6, p. 67. Let us emphasize that we can choose the constants uniformly with respect to m.

Integrating these expressions over Γ, applying the Fubini–Tonelli theorem, and then adding the resulting equivalences, we conclude that

$$\int_I \int_\Gamma |u(t,1,\theta)|^2 \, d\Gamma \, dt \asymp \sum_m \sum_k R_{m,k}(1)^2 \int_\Gamma \left| H^+_{m,k}(\theta) \right|^2 + \left| H^-_{m,k}(\theta) \right|^2 \, d\Gamma.$$

We obtain in a completely similar way that

$$\int_I \int_\Gamma |u'(t,1,\theta)|^2 \, d\Gamma \, dt$$

$$\asymp \sum_m \sum_k |\omega_{m,k}|^2 R_{m,k}(1)^2 \int_\Gamma \left| H^+_{m,k}(\theta) \right|^2 + \left| H^-_{m,k}(\theta) \right|^2 \, d\Gamma,$$

so that finally,

$$\int_I \int_\Gamma |u(t,1,\theta)|^2 + |u'(t,1,\theta)|^2 \, d\Gamma \, dt$$

$$\asymp \sum_m \sum_k \beta_{m,k} \int_\Gamma \left| H^+_{m,k}(\theta) \right|^2 + \left| H^-_{m,k}(\theta) \right|^2 \, d\Gamma \quad (7.21)$$

with

$$\beta_{m,k} := \left(1 + |\omega_{m,k}|^2 \right) R_{m,k}(1)^2. \quad (7.22)$$

These relations remain valid if some (only a finite number) of the exponents $\omega_{m,k}$ are equal to zero, which can be shown by a usual modification of the computation.

Comparing (7.19)–(7.20) and (7.21)–(7.22), we see that the inverse inequality (7.2) follows because

$$1 + c^2_{m,k} - m^2 \leq 1 + c^2_{m,k} \asymp 1 + |\omega_{m,k}|^2,$$

so that

$$\alpha_{m,k} \leq c\beta_{m,k}$$

with a suitable constant c.

On the other hand, the direct inequality (7.3) fails in general because we do not have

$$\beta_{m,k} \leq c' \alpha_{m,k}$$

with any constant c', independent of m and k. Indeed, it follows from part (d) of Lemma 7.5 (p. 132) that

$$\frac{\alpha_{m,1}}{\beta_{m,1}} \asymp \frac{1 + c_{m,1}^2 - m^2}{1 + c_{m,1}^2 + a} \to 0$$

as $m \to \infty$.

Remark. Using the strengthening of part (d) of Proposition 7.8, mentioned on page 135, one can establish a modified direct inequality by applying stronger Sobolev norms; see, e.g., [65], where this is done for a Petrovsky system, but the method is the same.

7.5 Observability of a Petrovsky System in a Ball

The proof given in the preceding section easily adapts to the case of the Petrovsky system considered in Section 3.6 (p. 50):

$$\begin{cases} u'' + \Delta^2 u + au = 0 & \text{in} \quad \mathbb{R} \times \Omega, \\ \partial_\nu u = \partial_\nu \Delta u = 0 & \text{on} \quad \mathbb{R} \times \Gamma, \\ u(0) = u_0 \quad \text{and} \quad u'(0) = u_1 & \text{in} \quad \Omega. \end{cases} \tag{7.23}$$

We recall from Proposition 3.6 (p. 51) that this problem is well-posed for $u_0 \in H^2(\Omega)$ and $u_0 \in L^2(\Omega)$ such that $\partial_\nu u_0 = 0$ on Γ. As usual, we consider only solutions with initial data belonging to the dense subspace $\mathcal{Z} := Z \times Z$, where Z denotes the linear hull of the eigenfunctions of $-\Delta$ with homogeneous Neumann boundary conditon in Ω.

Defining the energy by the formula

$$E(t) := \frac{1}{2} \int_\Omega |u(t,x)|^2 + |\Delta u(t,x)|^2 + |u'(t,x)|^2 \ dx, \quad t \in \mathbb{R},$$

it is natural to conjecture that

$$\int_I \int_\Gamma |u|^2 + |u'|^2 \ d\Gamma \ dt \asymp E(0)$$

for every interval I, because this estimate has already been established for the one-dimensional case in Proposition 5.1 (p. 84). However, this relation is inexact in several dimensions:

Proposition 7.11. *Let Ω be an open ball of radius R in $\mathbb{R}^N$ with $N \geq 2$.*

(a) Given any interval I, the inverse inequality

$$E(0) \leq c \int_I \int_\Gamma |u|^2 + |u'|^2 \ d\Gamma \ dt$$

holds for all solutions with $(u_0, u_1) \in \mathcal{Z}$, with a constant c depending on I.

(b) The direct inequality

$$\int_I \int_\Gamma |u|^2 + |u'|^2 \; d\Gamma \; dt \leq cE \tag{7.24}$$

can fail for any interval $|I|$.

Remarks.

- For $a = 0$ this result was first established in [65].
- Our proof will also show that if I_1 and I_2 are two intervals of length $> 2R$, then the corresponding observations are equivalent:

$$\int_{I_1} \int_\Gamma |u|^2 + |u'|^2 \; d\Gamma \; dt \asymp \int_{I_2} \int_\Gamma |u|^2 + |u'|^2 \; d\Gamma \; dt.$$

This property enables us to apply the abstract stabilization Theorem 2.14 (p. 24); see [70] for the details.

- The following proof of Proposition 7.11 can be easily adapted to the case of Dirichlet conditions. In this way we can get a new proof of part (b) of Proposition 6.10 (p. 111), but only in the special case of the balls. In this case the multiplier method, and also those of Carleman estimates and of microlocal analysis, are more efficient.
- As we have already remarked at the end of the preceding section, the direct inequality (7.24) holds if we introduce a stronger energy, corresponding to stronger Sobolev norms; see, e.g., [65].

Proof. According to Proposition 3.6 (p. 51), the solutions of (7.23) are given by the same formula (7.18); we have only to modify the definition of $\omega_{m,k}$ by this time setting

$$\omega_{m,k} := c_{m,k}^2 + a.$$

With this change, the proofs of the estimates (7.19) and (7.20) remain valid. Moreover, the proof of (7.20) remains valid for every (arbitrarily short) interval I. This follows from the fact that since $c_{m,1} \to \infty$ by part (d) of Proposition 7.9 (p. 137), parts (b) and (c) of the same proposition imply that

$$\inf_k \omega_{m,k+1} - \omega_{m,k} = \inf_k c_{m,k+1}^2 - c_{m,k}^2 \to \infty$$

as $m \to \infty$. Thus for any fixed interval I we may apply Ingham's theorem for all but finitely many indices m, which ensures the uniformity of the constants again. We leave the details to the reader.

7.6 Spherical Membranes and Plates

In this section we consider the problems

$$\begin{cases} u'' - \Delta u + au = 0 & \text{in } \mathbb{R} \times \Gamma, \\ u(0) = u_0 \quad \text{and} \quad u'(0) = u_1 & \text{in } \Gamma, \end{cases} \tag{7.25}$$

and

$$\begin{cases} u'' + \Delta^2 u + au = 0 & \text{in } \mathbb{R} \times \Gamma, \\ u(0) = u_0 \quad \text{and} \quad u'(0) = u_1 & \text{in } \Gamma, \end{cases} \tag{7.26}$$

where Γ is an N-dimensional sphere, i.e., the boundary of a ball in $\mathbb{R}^{N+1}$, and Δ denotes the Laplace–Beltrami operator on Γ. They describe the small transversal vibrations of spherical membranes and plates.

We study the internal observability of these problems by observing the solutions on some given *open* subset Γ_0 of Γ.

We may assume by scaling that Γ is the unit sphere of $\mathbb{R}^{N+1}$. Then we recall from Lemma 7.3 (p. 130) that $L^2(\Gamma)$ is the orthogonal direct sum of the finite-dimensional subspaces $\mathcal{S}_m$ formed by the spherical harmonics of order m for $m = 0, 1, \ldots$. Furthermore, we recall, e.g., from [127] that every $\mathcal{S}_m$ is an eigenspace of $-\Delta$ with the eigenvalue $\gamma_m = m(m + N - 1)$. Let us denote by Z the linear hull of the spherical harmonics, and for every real number s, let us denote by D^s the completion of Z with respect to the Euclidean norm

$$\left\| \sum_m a_m e_m \right\|_s^2 := \sum_m (1 + \gamma_m)^s |a_m|^2,$$

where e_m belongs to $\mathcal{S}_m$. Then, motivated by the works of Haraux [49] and Lions [97], pp. 405–407, it is natural for us to investigate the validity of the estimates

$$\|u_0\|_0^2 + \|u_1\|_{-1}^2 \asymp \int_I \int_{\Gamma_0} |u|^2 \, d\Gamma \, dt \tag{7.27}$$

for the solutions of (7.25), and the validity of the estimates

$$\|u_0\|_0^2 + \|u_1\|_{-2}^2 \asymp \int_I \int_{\Gamma_0} |u|^2 \, d\Gamma \, dt \tag{7.28}$$

for the solutions of (7.26).

Concerning the wave equation, the following result is a special case of a theorem of Rauch and Taylor [117], proved by microlocal analysis:

Proposition 7.12. *Let I be an interval of length $|I| > 2\pi$. Then the solutions of (7.25) satisfy the estimates (7.27) if and only if every great circle of Γ meets Γ_0.*

The purpose of this section is to establish a similar result for the plate model:

Proposition 7.13. *Let I be an arbitrary interval. Then the solutions of (7.26) satisfy the estimates (7.28) if and only if every great circle of Γ meets Γ_0.*

Remark. If we consider a sphere Γ of radius R, then the condition $T > 2\pi$ is replaced by $T > 2\pi R$ in Proposition 7.12, while Proposition 7.13 remains valid in the same form.

Instead of establishing Proposition 7.13 directly, we shall deduce it from Proposition 7.12, by reformulating the estimates (7.27) and (7.28) in terms of the spherical harmonics.

Lemma 7.14. *Let I be an interval of length $|I| > 2\pi$. Then the solutions of (7.25) satisfy the estimates (7.27) if and only if*

$$\int_\Gamma |e|^2 \, d\Gamma \asymp \int_{\Gamma_0} |e|^2 \, d\Gamma \tag{7.29}$$

for all spherical harmonics e.

Proof. Set $\omega_m := \sqrt{\gamma_m + a}$. If $e \in \mathcal{S}_m$, then $u(t,x) := e^{i\omega_m t}e(x)$ is a solution of (7.25). Applying (7.27), we obtain (7.29). Conversely, the general solution of (7.25) is given by the formula

$$u(t,x) = \sum_{m=0}^{\infty} e^{i\omega_m t} e_m^+(x) + e^{-i\omega_m t} e_m^-(x)$$

with suitable coefficients $e_m^+, e_m^- \in \mathcal{S}_m$, depending on the initial data. (It is sufficient to consider finite sums, as usual.) The family $\{\pm\sqrt{\gamma_m}\}$ satisfies the hypotheses of Theorem 4.6 with $\gamma' = 1$ (p. 67), so that

$$\int_I |u(t,x)|^2 \, d\Gamma \, dt \asymp \sum_{m=0}^{\infty} |e_m^+(x)|^2 + |e_m^-(x)|^2$$

for all $x \in \Gamma$, with corresponding constants independent of x. Integrating over Γ_0 and using Fubini's theorem, we obtain that

$$\int_0^T \int_{\Gamma_0} |u|^2 \, d\Gamma \, dt \asymp \sum_{m=0}^{\infty} \int_{\Gamma_0} |e_m^+|^2 + |e_m^-|^2 \, d\Gamma.$$

In view of (7.29) we still have an equivalent expression if we integrate over Γ instead of Γ_0, and we conclude by observing that

$$\sum_{m=0}^{\infty} \int_\Gamma |e_m^+|^2 + |e_m^-|^2 \, d\Gamma \asymp \|u_0\|_0^2 + \|u_1\|_{-1}^2$$

by a direct computation.

Lemma 7.15. *Let I be an arbitrary interval. The solutions of (7.26) satisfy the estimates (7.28) if and only if there exists a constant c such that the relations (7.29) are satisfied.*

Proof. We repeat the proof of the preceding lemma with three changes:

- we set $\omega_m := \sqrt{\gamma_m^2 + a}$;
- we apply Theorem 4.6 with $\gamma' = 0$;
- $\|u_1\|_{-1}^2$ is changed to $\|u_1\|_{-2}^2$.

Using Lemmas 7.14 and 7.15, Proposition 7.13 follows at once from Proposition 7.12.

7.7 Another Spherical Membrane

Spherical membranes of opening angle $\theta_0 = \pi/2$ were studied in [104]. It was shown in [42] that this system is not controllable; this result was generalized to systems of mixed order in [43].

In this section we deal with the *partial* exact controllability of spherical membranes of opening angle $0 < \theta_0 < \pi$, with a hole of opening angle $0 < \theta_1 < \theta_0$.

We denote by $u(t,\theta)$ and $w(t,\theta)$ the meridional and radial displacements, respectively. According to the linear shell theory of Love and Koiter [105], the vibration of this membrane is modeled by the following system:

$$
\begin{cases}
u_{tt} - \mathcal{L}u - (1-\nu)u + (1+\nu)w' = 0 & \text{in} \quad \mathbb{R} \times (\theta_1, \theta_0), \\
w_{tt} - \frac{1+\nu}{\sin\theta}(u\sin\theta)' + 2(1+\nu)w = 0 & \text{in} \quad \mathbb{R} \times (\theta_1, \theta_0), \\
u(t,\theta_1) = u(t,\theta_0) = 0 & \text{for} \quad t \in \mathbb{R}, \\
u(0,\theta) = u_0(\theta), \quad u_t(0,\theta) = u_1(\theta) & \text{for} \quad \theta \in (\theta_1, \theta_0), \\
w(0,\theta) = w_0(\theta), \quad w_t(0,\theta) = w_1(\theta) & \text{for} \quad \theta \in (\theta_1, \theta_0).
\end{cases} \tag{7.30}
$$

Here $'$ and the subscript stand for the derivatives with respect to θ and t, respectively, ν is a given constant satisfying $0 < \nu < 1/2$, and the operator $\mathcal{L}$ is defined by the following formula:

$$
\mathcal{L}v := v'' + v'\cos\theta - \frac{v}{\sin^2\theta}.
$$

One can show by applying the method of Chapter 3 that this problem is well-posed for

$$
(u_0, w_0, u_1, w_1) \in H_0^1(\theta_1, \theta_0) \times L^2(\theta_1, \theta_0) \times L^2(\theta_1, \theta_0) \times L^2(\theta_1, \theta_0).
$$

Furthermore, if (u, v) is a solution with $w_0 = w_1 = 0$, then we have

$$u(t,\theta) = \sum_{k=1}^{\infty} \left(a_k e^{i\omega_k^+ t} + a_{-k} e^{-i\omega_k^+ t}\right) e_k(\theta) + \sum_{k=1}^{\infty} \left(b_k e^{i\omega_k^- t} + b_{-k} e^{-i\omega_k^- t}\right) e_k(\theta)$$

with suitable complex numbers $a_{\pm k}$, $b_{\pm k}$, depending on the initial data, where $e_1, e_2, \ldots$ is an orthonormal basis in $L^2(\theta_1, \theta_0)$, formed by eigenfunctions of $-\mathcal{L}$ in $H_0^1(\theta_1, \theta_0)$, with corresponding eigenvalues $\lambda_1, \lambda_2, \ldots$, and the exponents $\omega_k^{\pm}$ are given by the following formulae:

$$\omega_k^{\pm} = \sqrt{\frac{(\lambda_k + 1 + 3\nu) \pm \sqrt{(\lambda_k + 1 + 3\nu)^2 - 4(1 - \nu^2)(\lambda_k - 2)}}{2}}. \qquad (7.31)$$

The following observability estimate is similar to some earlier results obtained in [104]:

Proposition 7.16. *Let I be an interval of length $> 2(\theta_0 - \theta_1)$. Then the solutions of (7.30) with $w_0 = w_1 = 0$ satisfy the estimates*

$$\int_I |u'(t, \theta_0)|^2 \, dt \asymp \sum \sum_{k \in \mathbb{Z}^*} \left(|a_k|^2 + |b_k|^2\right) |e_k'(\theta_0)|^2.$$

Remark. Applying the spectral theory as described in the first chapter of Titchmarsh's book [131], we obtain the asymptotic formula

$$\sqrt{\lambda_k} = \frac{k\pi}{\theta_0 - \theta_1} + o(1)$$

as $k \to \infty$. Using the expression (7.31) of the exponents $\omega_k^{\pm}$, it follows that

$$\omega_{k+1}^+ - \omega_k^+ \to \frac{\pi}{\theta_0 - \theta_1}$$

and

$$\omega_k^- \to \sqrt{1 - \nu^2}.$$

In particular, the family $\{\omega_k^{\pm}\}$ does not satisfy the uniform gap condition of Ingham's theorem.

In view of the last remark, we cannot apply Ingham's theorem. However, we will be able to conclude by applying one of its variants, to be described in the next section, by the following observation:

Lemma 7.17. *There exists a convergent series $\sum_{k \in \mathbb{Z}^*} p_k$ of positive numbers such that the solutions of (7.30) with $w_0 = w_1 = 0$ satisfy the following relations:*

$$|b_k|^2 \le p_k |a_k|^2 \quad \text{for all} \quad k.$$

Proof. Setting

$$\beta_k^+ := \frac{(\lambda_k - 1 + \nu)[(\omega_k^-)^2 - (\lambda_k - 1 + \nu)] - \lambda_k(1 + \nu)^2}{\omega_k^+[(\omega_k^-)^2 - (\omega_k^+)^2]}$$

and

$$\beta_k^- := \frac{(\lambda_k - 1 + \nu)[(\omega_k^+)^2 - (\lambda_k - 1 + \nu)] - \lambda_k(1 + \nu)^2}{\omega_k^-[(\omega_k^-)^2 - (\omega_k^+)^2]}$$

for brevity, we see that a straightforward computation leads to the relations

$$a_k = \frac{\beta_k^+}{(\omega_k^+)^2}\left(\rho_k\omega_k^+ + i\sigma_k\right) \quad \text{and} \quad b_k = \frac{\beta_k^-}{(\omega_k^-)^2}\left(\rho_k\omega_k^- - i\sigma_k\right)$$

with suitable *real* numbers ρ_k and σ_k, $k \in \mathbb{Z}^*$.

Using (7.31), a straightforward computation[3] leads to the following asymptotic estimates as $k \to \pm\infty$:

$$\omega_k^+ \asymp k,$$
$$\omega_k^- \asymp 1,$$
$$\beta_k^+ \asymp k,$$
$$\beta_k^- \asymp 1/k^2.$$

Since ρ_k and σ_k are real numbers, it follows that

$$\frac{\rho_k\omega_k^- - i\sigma_k}{\rho_k\omega_k^+ + i\sigma_k}$$

is bounded with respect to k and therefore

$$\frac{b_k}{a_k} = O(1/k).$$

We may thus choose $p_k = c/k^2$ with a sufficiently large constant c.

7.8 A Variant of Ingham's Theorem

We prove a variant of Ingham's theorem, whose application completes the proof of Proposition 7.16 in the preceding section. Although the result was obtained in [104], here the proof is given in a shorter form. Let $(\omega_k^+)_{k \in K}$ and $(\omega_k^-)_{k \in K}$ be two families of real numbers, $\gamma > 0$ a positive number, and $\sum_{k \in K} p_k$ a convergent series of positive numbers. Assume that

$$|\omega_k^+ - \omega_n^+| \geq \gamma \quad \text{for all} \quad k \neq n, \tag{7.32}$$
$$|\omega_k^+ - \omega_n^-| \geq \gamma \quad \text{for all} \quad k \text{ and } n, \tag{7.33}$$

---

[3]See [104] for the details.

and

$$\text{all elements of the family } (\omega_k^-)_{k \in K} \text{ are isolated.} \qquad (7.34)$$

Then the formula

$$\gamma' := \sup_{A \subset K} \inf_{\substack{k,n \in K \setminus A \\ k \neq n}} \{|\omega_k^+ - \omega_n^+|, |\omega_k^+ - \omega_n^-|\},$$

where A runs over the *finite* subsets of K, defines a number $\gamma' \geq \gamma$.

Theorem 7.18. *Assume (7.32)–(7.34). Then for every bounded interval I of length $|I| > \pi/\gamma'$, the estimates*

$$\int_I |x(t)|^2 \, dt \asymp \sum_{k \in K} \left(|x_k^+|^2 + |x_k^-|^2 \right) \qquad (7.35)$$

are satisfied for all sums

$$x(t) = \sum_{k \in K} x_k^+ e^{i\omega_k^+ t} + x_k^- e^{i\omega_k^- t}$$

with square-summable complex coefficients x_k^+, x_k^- such that

$$|x_k^-|^2 \leq p_k |x_k^+|^2 \quad \text{for all} \quad k. \qquad (7.36)$$

Remark. As usual, the *direct* part of (7.35) holds for every interval I.

Proof. As in the proof of Ingham's theorem (p. 62), if $G : \mathbb{R} \to \mathbb{R}$ is a continuous even function having a compact support and $g : \mathbb{R} \to \mathbb{R}$ is its Fourier transform, then

$$\frac{1}{2\pi} \int_{-\infty}^{\infty} g(t)|x(t)|^2 \, dt = \sum_{k,n \in K} \left(G(\omega_k^+ - \omega_n^+)x_k^+ \overline{x_n^+} + G(\omega_k^- - \omega_n^-)x_k^- \overline{x_n^-} \right)$$

$$+ \sum_{k,n \in K} \left(G(\omega_k^+ - \omega_n^-)x_k^+ \overline{x_n^-} + G(\omega_k^- - \omega_n^+)x_k^- \overline{x_n^+} \right).$$

If, moreover, G vanishes outside the interval $(-\pi, \pi)$, then the last sum vanishes by (7.33), while the first one reduces to

$$\left(G(0) \sum_{k \in K} |x_k^+|^2 \right) + \left(\sum_{k,n \in K} G(\omega_k^- - \omega_n^-)x_k^- \overline{x_n^-} \right)$$

by (7.32), so that

$$\frac{1}{2\pi} \int_{-\infty}^{\infty} g(t)|x(t)|^2 \, dt$$

$$= \left(G(0) \sum_{k \in K} |x_k^+|^2 \right) + \left(\sum_{k,n \in K} G(\omega_k^- - \omega_n^-)x_k^- \overline{x_n^-} \right). \qquad (7.37)$$

Applying Theorem 4.6 (p. 67),[4] we see that it suffices to establish the estimates (7.35) for sums over $K \backslash A$ instead of K, where A is an arbitrarily chosen finite subset of K. Choose a number $\gamma'' < \gamma'$ such that $|I| > 2\pi/\gamma''$ and then choose A such that the inequalities (7.32) and (7.33) are satisfied with γ'' instead of γ, and such that

$$\sum_{k \in K} p_k < \varepsilon$$

for a small $\varepsilon > 0$, to be chosen later. We may assume by a scaling argument that $\gamma'' = \pi$.

It is sufficient to prove the inverse inequality for intervals of the form $(-R, R)$ with $R > 1$. Choosing the same functions G and g as in the proof of Ingham's theorem, we have the following:

- $g \leq 0$ outside $(-R, R)$,
- $G(0) > 0$,
- G has finite maximum μ on $\mathbb{R}$.

Therefore, using (7.36), the last sum in (7.37) can be majorized as follows:

$$\left| \sum_{k,n \in K} G(\omega_k^- - \omega_n^-) x_k^- \overline{x_n^-} \right| \leq \mu \sum_{k,n \in K} |x_k^-| |x_n^-|$$

$$\leq \mu \sum_{k,n \in K} \left(\sqrt{p_n} |x_k^+| \right) \left(\sqrt{p_k} |x_n^+| \right)$$

$$\leq \frac{\mu}{2} \sum_{k,n \in K} \left(p_n |x_k^+|^2 + p_k |x_n^+|^2 \right)$$

$$= \mu \left(\sum_{k \in K} p_k \right) \left(\sum_{k \in K} |x_k^+|^2 \right)$$

$$\leq \mu \varepsilon \sum_{k \in K} |x_k^+|^2.$$

If we choose $\varepsilon < G(0)/\mu$ at the beginning, then using this estimate, we deduce from (7.37) the inequality

$$\sum_{k \in K} |x_k^+|^2 \leq \frac{\alpha}{2\pi (G(0) - \mu\varepsilon)} \int_{-\infty}^{\infty} g(t) |x(t)|^2 \, dt$$

with α denoting the maximum of g in $[-R, R]$. This proves the inverse inequality for the intervals $[-R, R]$. It remains valid by translation for all intervals of length > 2.

For the proof of the direct inequality, we choose again the same functions G and g as in the proof of Ingham's theorem. Then

---

[4]Here we use hypothesis (7.34).

- $g \geq 0$ on $\mathbb{R}$,
- $g \geq 1$ on some small interval $[-r, r]$,
- G attains it maximum in 0.

Therefore, we now deduce from (7.37) the following inequality:

$$\frac{\beta}{2\pi} \int_{-r}^{r} |x(t)|^2 \, dt \leq G(0)(1 + \varepsilon) \sum_{k \in K} |x_k^+|^2.$$

This proves the direct inequality for the interval $[-r, r]$. The general case follows by covering the interval I by finitely many translates of $[-r, r]$.

Multidimensional Ingham-Type Theorems

In this chapter we generalize Ingham's theorem to functions of several variables, i.e., to functions of the form

$$x(t) = \sum_{k \in K} x_k e^{i\omega_k \cdot t}, \quad t \in \mathbb{R}^N, \tag{8.1}$$

with complex coefficients x_k, where $(\omega_k)_{k \in K}$ is a given family of *vectors* in $\mathbb{R}^N$. We improve a classical theorem of Kahane: our results show an interesting connection between Ingham-type theorems and the spectral theory of the Laplacian operator. We also prove related optimal results by using other norms of the Euclidean space. We then apply these results in order to extend some surprising internal observability theorems of Haraux and Jaffard concerning rectangular plates to arbitrary spatial dimensions.

8.1 On a Theorem of Kahane

Let $(\omega_k)_{k \in K}$ be a family of *vectors* in $\mathbb{R}^N$. Given a Euclidean ball $B_R \subset \mathbb{R}^N$ of radius R, we are interested in the validity of the estimates

$$c_1 \sum_{k \in K} |x_k|^2 \le \int_{B_R} |x(t)|^2 \, dt \le c_2 \sum_{k \in K} |x_k|^2 \tag{8.2}$$

with suitable (strictly) positive constants c_1 and c_2, independent of the particular choice of the coefficients x_k.

Fix $1 \le p \le \infty$ arbitrarily and let us denote by B_r^p the ball of radius r in $\mathbb{R}^N$ with respect to the p-norm:

$$B_r^p := \{ x \in \mathbb{R}^N \; : \; \|x\|_p < r \}.$$

Here

$$\|x\|_p := \left(\sum_{i=1}^{N}|x_i|^p\right)^{1/p}$$

if $1 \le p < \infty$ and

$$\|x\|_\infty := \max\{|x_1|,\ldots,|x_N|\}$$

for $p = \infty$. Similarly to the scalar case, (8.2) cannot hold unless there exists a number $\gamma = \gamma_p > 0$ such that

$$\|\omega_k - \omega_n\|_p \ge \gamma \quad \text{for all} \quad k \neq n. \tag{8.3}$$

We are going to prove the following result:

Theorem 8.1. *Assume (8.3) and let us denote by μ_p the first eigenvalue of $-\Delta$ in the Sobolev space $H_0^1(B_{\gamma/2}^p)$. If $R > \sqrt{\mu_p}$, then all sums (8.1) with square-summable complex coefficients x_k satisfy the estimates (8.2).*

Remarks.

- For $N = 1$ this result reduces to Ingham's theorem (p. 59) because the first eigenfunction of $-\Delta$ in $H_0^1(-\gamma/2, \gamma/2)$ is (up to a multiplicative constant) the already familiar function $\cos(\pi x/\gamma)$, so that $\mu_p = \pi^2/\gamma^2$, and B_R is the interval $(-R, R)$ of length $2R$.[1]
- In the Euclidean case $p = 2$ the above result strenghtens an earlier theorem of Kahane [61], by providing a weaker condition on R for the validity of (8.3). In terms of the Bessel functions our condition can be reformulated in the form $R > 2\rho_N/\gamma$, where ρ_N denotes the first positive root of the Bessel function $J_{\frac{N}{2}-1}(x)$.
- The theorem was obtained in [7] for $p = 2$ and $p = \infty$, and then in [103] for the general case.

We prove this theorem by adapting Ingham's method given in Section 4.2 (p. 62): instead of estimating $\int_{B_R}|x(t)|^2\,dt$ directly, we estimate $\int_{\mathbb{R}^N} g(t)|x(t)|^2\,dt$, where $g(t)$ is a carefully chosen weight function for which the exponential functions $e^{i\omega_k \cdot t}$ are mutually orthogonal. We proceed in four steps.

First step: translation invariance. If (8.2) holds for some ball, then it also holds for every other ball of the same radius, with the same constants c_1 and c_2. This follows by a simple change of variables. Hence it suffices to consider (Euclidean) balls B_R centered at the origin.

Second step: Fourier transfom. If $G \in H_0^1(B_\gamma^p)$ and G is continuous, then extending it by zero outside B_γ^p and introducing its Fourier transform

$$g(t) := \int_{B_\gamma^p} G(x)e^{-ix \cdot t}\,dx, \quad t \in \mathbb{R}^N,$$

---

[1]Of course, in the one-dimensional case all p-norms are the same.

we have

$$\int_{\mathbb{R}^N} g(t)|x(t)|^2 \, dt = (2\pi)^N G(0) \sum_{k \in K} |x_k|^2. \tag{8.4}$$

Indeed, this follows from the identity

$$\int_{\mathbb{R}^N} g(t)|x(t)|^2 \, dt = (2\pi)^N \sum_{k,n \in K} x_k \overline{x_n} G(\omega_k - \omega_n),$$

because as a consequence of (8.3), $G(\omega_k - \omega_n) = 0$ whenever $k \neq n$.

Third step: proof of the second estimate in (8.2). Let us denote by H the eigenfunction of $-\Delta$ in $H_0^1(B_{\gamma/2}^p)$, corresponding to the first eigenvalue μ. Multiplying by -1 if necessary, we may assume that $H > 0$ in $B_{\gamma/2}^p$. Extending by zero outside this ball, we obtain a continuous function on $\mathbb{R}^N$, still denoted by H.

Let us denote by h the Fourier transform of H. Since the function H is radial and therefore even, for[2] $\|t\|_q \leq \pi/\gamma$ we have

$$h(t) = \int_{B_{\gamma/2}} H(x) e^{-ix \cdot t} \, dx = \int_{B_{\gamma/2}} H(x) \cos x \cdot t \, dx > 0$$

because for $\|x\|_p < \gamma/2$ and $\|t\|_q \leq \pi/\gamma$ we have $|x \cdot t| < \gamma/2$, and therefore $\cos x \cdot t > 0$. Since h is continuous, it has thus a positive minimum on the compact set $B := \{t \in \mathbb{R}^N \ : \ \|t\|_q \leq \pi/\gamma\}$.

With $G := H * H$, it follows that

$$G \in H_0^1(B_\gamma^p),$$
$$g = |h|^2 \geq 0 \quad \text{in} \quad \mathbb{R}^N,$$

and

$$g \quad \text{has a positive lower bound } \beta \text{ in } B.$$

Therefore, using (8.4) we have

$$\beta \int_B |x(t)|^2 \, dt \leq \int_{\mathbb{R}^N} g(t)|x(t)|^2 \, dt = (2\pi)^N G(0) \sum_{k \in K} |x_k|^2,$$

proving the second inequality of (8.2) for B instead of B_R, with $c_2 = (2\pi)^N G(0)/\beta$. The general case follows by covering B_R with a finite number of translates of B and applying the triangle inequality.

Fourth step: proof of the first estimate in (8.2). Consider the same function H as before. Now setting $G := (R^2 + \Delta)(H * H)$ and denoting its Fourier transform by g, we have

---

[2]Here we denote by q the conjugate exponent of p.

$$G \in H_0^1(B_\gamma^p),$$
$$g(t) = (R^2 - |t|^2)|h(t)|^2 \leq 0 \quad \text{if} \quad |t| \geq R.$$

Hence g is bounded from above in $\mathbb{R}^N$ by some constant α. Using (8.4) we obtain that

$$(2\pi)^N G(0) \sum_{k \in K} |x_k|^2 = \int_{\mathbb{R}^N} g(t)|x(t)|^2 \, dt \leq \alpha \int_{B_R} |x(t)|^2 \, dt.$$

This yields the first inequality of (8.2) with $c_1 = (2\pi)^N G(0)/\alpha$, provided that $G(0) > 0$.

This last inequality follows from our assumption $R > \sqrt{\mu_p}$ and from the variational characterization of the first eigenvalue μ of the operator $-\Delta$ in $H_0^1(B_{\gamma/2}^p)$:

$$G(0) = \int_{B_{\gamma/2}^p} R^2 H^2 - |\nabla H|^2 \, dx = (R^2 - \mu_p) \int_{B_{\gamma/2}^p} H^2 \, dx > 0.$$

8.2 On the Optimality of Theorem 8.1

The optimality of the condition $R > \sqrt{\mu_p}$ in Theorem 8.1 remains a very interesting open question in the general case. In this section we establish the optimality in two particular cases.

As in the preceding section, let $(\omega_k)_{k \in K}$ be a family of *vectors* in $\mathbb{R}^N$, satisfying for some $1 \leq p \leq \infty$ and $\gamma > 0$ the gap condition (8.3). We are going to prove the following theorem:

Theorem 8.2. *Assume* (8.3).

(a) *The inverse inequality in* (8.2) *can fail if* $p = \infty$ *and* $R < \sqrt{\mu_\infty}$.

(b) *The inverse inequality in* (8.2) *can also fail if* $N = 2$, $p = 1$, *and* $R < \sqrt{\mu_1}$.

Remarks.

- Part (a) was proved in [7] in an even more general form, by considering also functions of the form

$$x(t) = \sum_{k \in K} \sum_{|j| < M} x_{k,j} t^j e^{i\omega_k \cdot t}, \quad t \in \mathbb{R}^N,$$

where M is a given positive integer, $j = (j_1, \ldots, j_N)$ is a multi-index, and

$$|j| = j_1 + \cdots + j_N, \quad t^j = t_1^{j_1} \cdots t_N^{j_N}.$$

- Part (b) seems to be new.

A simple scaling argument shows that it is sufficient to establish the above results for an arbitrarily chosen $\gamma > 0$.

Proof of part (a). Let the exponents ω_k run over the set $\mathbb{Z}^N$ of points of $\mathbb{R}^N$ all of whose coordinates are integers. Then (8.3) is satisfied with $p = \infty$ and $\gamma = 1$.

Fix an arbitrarily small number $0 < \varepsilon < \pi$ and set (see Figure 8.1)

$$x_\varepsilon(t) := \begin{cases} 1 & \text{if dist } (t, 2\pi\mathbb{Z}^N) < \varepsilon, \\ 0 & \text{otherwise,} \end{cases}$$

where the distance is taken in the usual Euclidean sense. The function x_ε is locally square-summable and 2-periodic in each variable. Developing it into an N-fold trigonometric Fourier series, we thus have

$$x_\varepsilon(t) = \sum_{k \in \mathbb{Z}^N} x_k e^{ik \cdot t}$$

with square-summable coefficients x_k. Since x_ε does not vanish identically, there are nonzero coefficients. Since x_ε vanishes identically in the ball of center $(\pi, \ldots, \pi)$ and radius $\sqrt{N}\pi - \varepsilon$, we conclude that the first inequality of (8.2) cannot hold unless $R > \sqrt{N}\pi - \varepsilon$. Since ε can be chosen arbitrarily small, we complete the proof by observing that $\sqrt{\mu_\infty} = \sqrt{N}\pi$ because the first eigenfunction of $-\Delta$ in $H_0^1(B_{1/2}^\infty)$ is given by the formula

$$H(x) = \cos(\pi x_1) \cdots \cos(\pi x_N),$$

so that $-\Delta H = N\pi^2 H$.

Proof of part (b). Consider the function x_ε introduced above. If $N = 2$, then setting

$$n = (k_1 + k_2, k_1 - k_2) \quad \text{and} \quad s = \left(\frac{t_1 + t_2}{2}, \frac{t_1 - t_2}{2} \right),$$

we have $k \cdot t = n \cdot s$, so that we may rewrite the Fourier expansion of x_ε in the new variable s in the form

$$x_\varepsilon(t(s)) = \sum_n x_{k(n)} e^{in \cdot s},$$

where $n = (n_1, n_2)$ runs over the elements of $\mathbb{Z}^2$ for which $n_1 + n_2$ is even. Since $\|t\|_2 = \sqrt{2}\|s\|_2$ and therefore

$$\int_{B_{R\sqrt{2}}} |x_\varepsilon(t)|^2 \, dt = 2 \int_{B_R} |x_\varepsilon(t(s))|^2 \, ds,$$

it follows from part (a) that the first inequality of (8.2) cannot hold unless $R\sqrt{2} \geq \sqrt{N}\pi$ with $N = 2$; i.e. $R \geq \pi$. We conclude by observing that the set of exponents n satisfies $\gamma = 2$ for $p = 1$, and that $\mu_1 = \pi^2$ in the present case because $B_{\gamma/2}^1$ is a square of side $\sqrt{2}$, so that the first eigenvalue of $-\Delta$ in B_1^1 is equal to π^2.

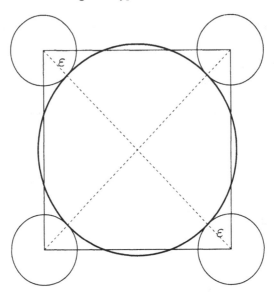

Fig. 8.1. Proof of part (b)

8.3 A Variant of Haraux's Theorem

In this section we extend Theorem 4.5 (p. 67) to the multidimensional case. Let $(\lambda_k)_{k \in K}$ be a family of vectors in $\mathbb{C}^N$ satisfying

$$\sup |\Re \lambda_k| < \infty,$$

and for some $k_0 \in K$ the *gap condition*

$$\gamma_0 := \inf_{k \neq k_0} \|\lambda_k - \lambda_{k_0}\|_2 > 0.$$

Proposition 8.3. *Assume that for some ball $B_0 \subset \mathbb{R}^N$ we have*

$$\int_{B_0} |x(t)|^2 \, dt \asymp \sum_{k \in K \setminus \{k_0\}} |x_k|^2 \tag{8.5}$$

for all finite *sums of the form*

$$x(t) = \sum_{k \in K \setminus \{k_0\}} x_k e^{\lambda_k \cdot t} \tag{8.6}$$

with complex coefficients x_k. Then for every ball B having a strictly bigger radius we also have

$$\int_B |x(t)|^2 \, dt \asymp \sum_{k \in K} |x_k|^2 \tag{8.7}$$

for all sums of the form

$$x(t) = \sum_{k \in K} x_k e^{\lambda_k \cdot t} \tag{8.8}$$

with square-summable complex coefficients x_k. The corresponding constants depend only on the balls B_0 and B.

Remark. We could also establish a more general result, similar to Theorem 6.2 (p. 93) by allowing more general terms $t^m e^{\lambda_k \cdot t}$ in the sums where m is a multi-index. We consider only a less general situation, in order not to make the book too long.

The proof is a straightforward adaptation of that given in Section 4.4 (p. 69), so we merely indicate the main steps. As in the one-dimensional case, it suffices to establish (8.7) for finite sums: the general case then follows by density. By rearranging the terms if necessary, we may assume that K is the set of nonnegative integers and $k_0 = 0$.

Proof of the *direct* part of (8.7). We can repeat the corresponding arguments in Section 4.4. First, the estimates (8.5) remain valid for every translate of the ball B_0. Then the direct part of (8.7) can be established on all these translated balls by an elementary argument using the triangle inequality. Finally, the direct part of (8.7) is proved for every ball by covering it by finitely many translates of B_0 and using the triangle inequality again.

Proof of the *inverse* part of (8.7). Given a ball B whose radius is bigger than that of B_0, choose a positive number δ that is smaller than the difference of the radii. Since the estimates (8.5) are valid for every translate of the ball B_0, we may assume that the balls B_0 and B are concentric.

For x given by (8.8), the formula

$$y(t) := x(t) - \frac{1}{V_\delta} \int_{B_\delta} e^{-\lambda_0 \cdot s} x(t+s) \, ds,$$

where V_δ denotes the volume of the ball

$$B_\delta := \{s \in \mathbb{R}^N \ : \ |s| < \delta\},$$

defines a function y of the form (8.6): an easy computation shows that

$$y(t) = \sum_{k=1}^\infty (1 - g(\lambda_k - \lambda_0)) x_k e^{\lambda_k \cdot t} =: \sum_{k=1}^\infty y_k e^{\lambda_k \cdot t}$$

with

$$g(\lambda) := \frac{1}{V_\delta} \int_{B_\delta} e^{\lambda \cdot s} \, ds.$$

If λ is a nonzero, purely imaginary vector, then $|g(\lambda)| < 1$, and $g(\lambda) \to 0$ as $|\lambda| \to \infty$. If all vectors $\lambda_k - \lambda_0$ are purely imaginary, then we may conclude by the gap condition $\gamma_0 > 0$ that

$$\varepsilon := \inf_{k \geq 1} |1 - g(\lambda_k - \lambda_0)|^2 > 0.$$

The same conclusion may be obtained in the general case, too, by slightly changing δ if necessary.

Then using the assumption (8.5) we have

$$\sum_{k=1}^{\infty} |x_k|^2 \leq \varepsilon^{-1} \sum_{k=1}^{\infty} |y_k|^2 \leq c_1 \int_{B_0} |y(t)|^2 \, dt \qquad (8.9)$$

with a suitable constant c_1. Furthermore,

$$|y(t)|^2 \leq 2|x(t)|^2 + 2 \left| \frac{1}{V_\delta} \int_{B_\delta} e^{-\lambda_0 \cdot s} x(t+s) \, ds \right|^2$$

$$\leq 2|x(t)|^2 + \frac{2e^{2|\Re\lambda_0|\delta}}{V_\delta} \int_{B_\delta} |x(t+s)|^2 \, ds$$

$$= 2|x(t)|^2 + \frac{2e^{2|\Re\lambda_0|\delta}}{V_\delta} \int_{B_\delta(t)} |x(s)|^2 \, ds,$$

so that

$$\int_{B_0} |y(t)|^2 \, dt \leq 2 \int_{B_0} |x(t)|^2 \, dt + \frac{2e^{2|\Re\lambda_0|\delta}}{V_\delta} \int_{B_0} \int_{B_\delta(t)} |x(s)|^2 \, ds \, dt$$

$$= 2 \int_{B_0} |x(t)|^2 \, dt + \frac{2e^{2|\Re\lambda_0|\delta}}{V_\delta} \int_{B} \int_{B_\delta(s) \cap B_0} |x(s)|^2 \, dt \, ds$$

$$\leq 2 \int_{B_0} |x(t)|^2 \, dt + 2e^{2|\Re\lambda_0|\delta} \int_{B} |x(s)|^2 \, ds$$

$$\leq \left(2 + 2e^{2|\Re\lambda_0|\delta} \right) \int_{B} |x(s)|^2 \, ds.$$

Combining this with (8.9), we conclude that

$$\sum_{k=1}^{\infty} |x_k|^2 \leq c_2 \int_{B} |x(s)|^2 \, ds \qquad (8.10)$$

with a suitable constant c_2. In order to recover (8.7) it remains to establish the estimate

$$|x_0|^2 \leq c_3 \int_{B} |x(s)|^2 \, ds \qquad (8.11)$$

with a suitable constant c_3. Using the decomposition (8.8), we have

$$|x_0|^2 \leq c_4 \int_{B} |x_0|^2 \, dt \leq 2c_4 \int_{B} |x(t)|^2 + \left| \sum_{k=1}^{\infty} x_k e^{\lambda_k \cdot t} \right|^2 \, dt$$

with some constant c_4. Since using our assumption (8.5) and then (8.10) we have

$$\int_B \Big| \sum_{k=1}^\infty x_k e^{\lambda_k \cdot t} \Big|^2 \, dt \le c_5 \sum_{k=1}^\infty |x_k|^2 \le 4c_2 c_5 \int_B |x(s)|^2 \, ds$$

with another constant c_5, we deduce from the preceding inequality that

$$|x_0|^2 \le c_4 \int_B |x_0|^2 \, dt \le 2c_4 \int_B |x(t)|^2 \, dt + 8c_2 c_4 c_5 \int_B |x(t)|^2.$$

Thus (8.11) holds with $c_3 := 2c_4 + 8c_2 c_4 c_5$.

8.4 A Weakening of Ingham's Condition

We have already given a weakening of Ingham's condition $|I| > 2\pi/\gamma$ in Theorem 4.6, p. 67. The following result of Kahane is stronger; however, the proof is more involved. The results established in this section will also be used in the following chapter.

Proposition 8.4. *Let $(\omega_k)_{k \in K}$ be a family of vectors in $\mathbb{R}^N$, satisfying the uniform gap condition. Let $K = K_1 \cup \cdots \cup K_m$ be a partition of K, and assume that the estimates*

$$\int_{B_{R_j}} \Big| \sum_{k \in K_j} x_k e^{i\omega_k \cdot t} \Big|^2 \, dt \asymp \sum_{k \in K_j} |x_k|^2, \quad j = 1, \ldots, m, \qquad (8.12)$$

hold for all finite sums with complex coefficients x_k, with suitable numbers R_j. Then we also have

$$\int_{B_R} \Big| \sum_{k \in K} x_k e^{i\omega_k \cdot t} \Big|^2 \, dt \asymp \sum_{k \in K} |x_k|^2 \qquad (8.13)$$

for every $R > R_1 + \cdots + R_m$, for all sums with square-summable complex coefficients.

As usual, we may assume by translation that all balls are centered at zero, and it suffices to establish (8.13) for finite sums.

If $R \le \min R_j$, then the direct part of (8.13) follows from (8.12) by using the elementary inequality

$$|z_1 + \cdots + z_m|^2 \le m\big(|z_1|^2 + \cdots + |z_m|^2\big).$$

If $R > \min R_j$, then the direct inequalities are obtained by covering B_R by finitely many small balls and applying the already proved inequality on each of them.

Turning to the proof of the inverse inequality, first we recall a basic result of functional analysis on *biorthogonal families*:

Lemma 8.5. *Let $(f_k)_{k \in K}$ be a family of vectors in a Hilbert space H. Assume that*

$$c_1 \sum_{k \in K} |x_k|^2 \leq \left\| \sum_{k \in K} x_k f_k \right\|^2 \leq c_2 \sum_{k \in K} |x_k|^2 \tag{8.14}$$

for all finite linear combinations of these vectors, with two positive constants c_1 and c_2. Then there exists another family $(y_k)_{k \in K}$ of vectors in H such that

$$(y_k, f_n) = \delta_{kn} \tag{8.15}$$

for all $k, n \in K$. Moreover, both families are bounded in H:

$$\|f_k\| \leq \sqrt{c_2} \quad and \quad \|y_k\| \leq 1/\sqrt{c_1}$$

for all k.

Proof. The inequalities $\|f_k\| \leq \sqrt{c_2}$ follow at once from (8.14). Let us denote by w_n the orthogonal projection of f_n onto the closed linear subspace H_m of H spanned by the remaining vectors f_k. Then we have

$$(f_n - w_n, f_k) = 0 \quad \text{for all} \quad k \neq n$$

and

$$(f_n - w_n, w_n) = 0,$$

whence

$$(f_n - w_n, f_n) = \|f_n - w_n\|^2.$$

Writing $f_n - w_n = \sum x_k f_k$, we have $x_n = 1$, so that $\|f_n - w_n\|^2 \geq c_1$ by (8.14). Therefore, the formula

$$y_n := \|f_n - w_n\|^{-2}(f_n - w_n)$$

defines a family satisfying (8.15) and bounded by $1/\sqrt{c_1}$.

Turning to the proof of the proposition, we may (and will) assume that $m = 2$: the general case then follows by induction on m. Furthermore, we will extend every function $\rho \in L^2(B)$ by zero to $\mathbb{R}^N$; equivalently, $\rho \in L^2(B)$ if $\rho \in L^2(\mathbb{R}^N)$ and $\rho = 0$ almost everywhere outside B. Setting henceforth $f_n(t) := e^{i\omega_n \cdot t}$ for brevity, we then have

$$(\rho, f_n)_{L^2(B)} = \widehat{\rho}(\omega_n)$$

for all n by the definition of the Fourier transform.

Fix a positive number r such that $R \geq R_1 + R_2 + 3r$.

Lemma 8.6. *There exists a bounded family $(\rho_k)_{k \in K}$ in $L^2(B_{R_1+R_2+r})$ satisfying*

$$\widehat{\rho}_k(\omega_n) = \delta_{kn}$$

for all $k, n \in K$.

Proof. As a consequence of hypothesis (8.12) we may apply the preceding lemma to the family $(f_k)_{k \in K_j}$ in $L^2(B_{R_j})$, $j = 1, 2$: there exists a bounded family $(\varphi_k)_{k \in K_j}$ in $L^2(B_{R_j})$ satisfying

$$\widehat{\varphi}_k(\omega_n) = \delta_{kn}$$

for all $k, n \in K_j$, $j = 1, 2$. Since B_{R_j} has finite volume, the family $(\varphi_k)_{k \in K_j}$ is also bounded in $L^1(B_{R_j})$ by Hölder's inequality.

Furthermore, for each fixed $k \in K_1$, as a consequence of Proposition 8.3 (p. 158) we may also apply the preceding lemma to the family $(f_n)_{n \in K_2 \cup \{k\}}$ in $L^2(B_{R_2+r})$. In particular, there exists a function $\psi_k \in L^2(B_{R_2+r})$ satisfying

$$\widehat{\psi}_k(\omega_k) = 1, \quad \text{and} \quad \widehat{\psi}_k(\omega_n) = 0 \quad \text{for all} \quad n \in K_2.$$

Similarly, for every $k \in K_2$ there exists a function $\psi_k \in L^2(B_{R_1+r})$ satisfying

$$\widehat{\psi}_k(\omega_k) = 1, \quad \text{and} \quad \widehat{\psi}_k(\omega_n) = 0 \quad \text{for all} \quad n \in K_1.$$

Note that by the uniform gap condition on the whole family $(\omega_k)_{k \in K}$, we may choose the same constants in all estimates provided by Proposition 8.3, so that the family $\{\psi_k : k \in K_1\}$ is bounded in $L^2(B_{R_2+r})$, and the family $\{\psi_k : k \in K_2\}$ is bounded in $L^2(B_{R_1+r})$. Since the family $(\varphi_k)_{k \in K_j}$ is bounded in $L^1(B_{R_j})$, the convolution product

$$\rho_k := \varphi_k * \psi_k$$

defines a bounded family $\{\rho_k : k \in K\}$ in $L^2(B_{R_1+R_2+r})$. (Here we use the elementary properties of the support of a convolution product and the fact that all balls are centered at the origin.)

This family has the required properties. Indeed, we have

$$\widehat{\rho}_k(\omega_k) = \widehat{\varphi}_k(\omega_k)\widehat{\psi}_k(\omega_k) = 1 \cdot 1 = 1$$

for all $k \in K$. It remains to show that if $k \neq n$, then $\widehat{\rho}_k(\omega_n) = 0$. This follows from the equality

$$\widehat{\rho}_k(\omega_n) = \widehat{\varphi}_k(\omega_n)\widehat{\psi}_k(\omega_n)$$

because one of the two factors on the right-hand side vanishes. Indeed, $\varphi_k(\omega_n) = 0$ if k and n belong to a same index set K_j, and $\psi_k(\omega_n) = 0$ otherwise.

Now we can establish the inverse inequality of the proposition:

Lemma 8.7. *We have*

$$\sum_{k \in K} |x_k|^2 \leq c \int_{B_R} |x(t)|^2 \, dt$$

for every finite sum

$$x(t) = \sum_{k \in K} x_k f_k(t) = \sum_{k \in K} x_k e^{i\omega_k \cdot t}$$

with complex coefficients x_k, where c is a constant independent of the particular choice of (x_k).

Proof. It is sufficient to construct a function $y \in L^2(B_R)$ satisfying

$$\sum_{k \in K} |x_k|^2 = (y, x)_{L^2(B_R)} \quad \text{and} \quad \|y\|_{L^2(B_R)}^2 \leq c \sum_{k \in K} |x_k|^2.$$

Then applying the Cauchy–Schwarz inequality, we have

$$\left(\sum_{k \in K} |x_k|^2 \right)^2 \leq \|y\|_{L^2(B_R)}^2 \|x\|_{L^2(B_R)}^2 \leq c \left(\sum_{k \in K} |x_k|^2 \right) \|x\|_{L^2(B_R)}^2,$$

and the required estimate follows.

For the construction of y we are going to use the functions H and h introduced in the proof of Theorem 8.1 (p. 154) but by reversing the role of the variables t and x. More precisely, let us denote by $H(t)$ the first eigenfunction of $-\Delta$ in $H_0^1(B_r)$, normalized so that

$$\int_{\mathbb{R}^N} H(t) \, dt = \int_{B_r} H(t) \, dt = 1,$$

and let us introduce its Fourier transform $h(\omega)$ by the formula

$$h(\omega) := \int_{\mathbb{R}^N} H(t) e^{-i\omega \cdot t} \, dt, \quad \omega \in \mathbb{R}^N.$$

Then $G := H * H \in H_0^1(B_{2r})$ is a bounded continuous function, whose Fourier transform

$$g(\omega) := \int_{\mathbb{R}^N} G(t) e^{-i\omega \cdot t} \, dt, \quad \omega \in \mathbb{R}^N,$$

is nonnegative because

$$g(\omega) = |h(\omega)|^2,$$

and we have

$$g(0) = |h(0)|^2 = 1$$

by the above normalization of H.

Let us also introduce the functions

$$G_n(t) := G(t) e^{i\omega_n \cdot t}, \quad t \in \mathbb{R}^N, \quad n \in K.$$

Their Fourier transforms

$$g_n(\omega) := \int_{\mathbb{R}^N} G_n(t) e^{-i\omega \cdot t} \, dt$$

satisfy the simple algebraic relations

$$g_n(\omega) = g(\omega - \omega_n)$$

because

$$g_n(\omega) = \int_{\mathbb{R}^N} G_n(t)e^{-i\omega \cdot t}\, dt = \int_{\mathbb{R}^N} G(t)e^{-i(\omega-\omega_n)\cdot t}\, dt = g(\omega - \omega_n).$$

Note that $G_n \in H_0^1(B_{2r})$ for every n.

Now for any given

$$x(t) = \sum_{k\in K} x_k f_k(t) = \sum_{k\in K} x_k e^{i\omega_k \cdot t}$$

we define a function $g(t)$ by the explicit formula

$$g(t) = \frac{1}{(2\pi)^N} \sum_{n\in K} x_n (\rho_n * K_n)(t).$$

Since ρ_n vanishes outside $B_{R_1+R_2+r}$, G_n vanishes outside B_{2r} and $R \geq R_1 + R_2 + 3r$, g belongs to $L^2(B_R)$.

A straightforward computation shows that

$$
\begin{aligned}
(y,x)_{L^2(B_R)} &= \frac{1}{(2\pi)^N} \sum_{n,k\in K} x_n \overline{x_k} \widehat{\rho}_n(\omega_k)\widehat{G}_n(\omega_k) \\
&= \sum_{n,k\in K} x_n \overline{x_k}\widehat{\rho}_n(\omega_k)g(\omega_k - \omega_n) \\
&= \sum_{k\in K} |x_k|^2
\end{aligned}
$$

because $\widehat{\rho}_n(\omega_k) = \delta_{nk}$ and $g(0) = 1$.

It remains to estimate the L^2 norm of y. Using Plancherel's equality we have

$$
\begin{aligned}
\|y\|_{L^2(B_R)}^2 &= \int_{\mathbb{R}^N} |y(t)|^2\, dt \\
&= (2\pi)^{-N} \int_{\mathbb{R}^N} |\widehat{y}(\omega)|^2\, d\omega \\
&= (2\pi)^{-N} \int_{\mathbb{R}^N} \left|\sum_{n\in K} x_n \widehat{\rho}_n(\omega)g_n(\omega)\right|^2 d\omega.
\end{aligned}
$$

Since the Fourier transform sends $L^1(\mathbb{R}^N)$ continuously into $L^\infty(\mathbb{R}^N)$ and since the family (ρ_n) is bounded in $L^2(B_R)$ and thus also in $L^1(\mathbb{R}^N)$, the family $(\widehat{\rho}_n)$ is uniformly bounded in $\mathbb{R}^N$ by some constant c_1. Since, moreover, $g_n \geq 0$ in $\mathbb{R}^N$, we deduce from the above equality the estimate

$$\|y\|^2_{L^2(B_R)} \le (2\pi)^{-N} c_1^2 \int_{\mathbb{R}^N} \Big| \sum_{n \in K} |x_n| g_n(\omega) \Big|^2 \, d\omega.$$

Using Plancherel's equality again, we see that this is equivalent to the inequality

$$\|y\|^2_{L^2(B_R)} \le c_1^2 \int_{\mathbb{R}^N} \Big| \sum_{n \in K} |x_n| G_n(t) \Big|^2 \, dt = \int_{\mathbb{R}^N} \Big| \sum_{n \in K} |x_n| G(t) e^{i\omega_n \cdot t} \Big|^2 \, dt.$$

Since the function $G(t)$ is bounded in B_{2r} and vanishes outside this ball, it follows that

$$\|y\|^2_{L^2(B_R)} \le c_2 \int_{B_{2r}} \Big| \sum_{n \in K} |x_n| e^{i\omega_n \cdot t} \Big|^2 \, dt$$

with another constant c_2. We conclude by recalling that the direct inequality has already been established, so that the right-hand side of the last inequality is majorized by a constant multiple of

$$\sum_{k \in K} |x_k|^2.$$

8.5 Internal Observability of Petrovsky Systems

Let us return to the system

$$\begin{cases} u'' + \Delta^2 u + au = 0 & \text{in} \quad \mathbb{R} \times \Omega, \\ u = \Delta u = 0 & \text{on} \quad \mathbb{R} \times \Gamma, \\ u(0) = u_0 \quad \text{and} \quad u'(0) = u_1 & \text{in} \quad \Omega, \end{cases} \tag{8.16}$$

of Section 3.5 (p. 48). We recall from Proposition 3.5 that this problem is well-posed for

$$u_0 \in H^2(\Omega) \cap H_0^1(\Omega) \quad \text{and} \quad u_1 \in L^2(\Omega),$$

and that the energy of the solutions is defined by the formula

$$E(t) := \frac{1}{2} \int_\Omega |u(t,x)|^2 + |\Delta u(t,x)|^2 + |u'(t,x)|^2 \, dx, \quad t \in \mathbb{R}.$$

Assuming that we can observe the solution only on some *open* subset Ω_0 of Ω and a time interval I, we are led to investigate the validity of the estimates

$$\int_I \int_{\Omega_0} |u'|^2 \, dx \, dt \asymp E(0). \tag{8.17}$$

They hold if $\Omega_0 = \Omega$. Indeed, writing[3]

---

[3]See Proposition 3.5, p. 48. We assume for simplicity that $\omega_m \ne 0$ for all m. The results remain valid without this assumption by the usual modifications of the proofs.

$$u(t,x) = \sum_{k=1}^{\infty} (a_k e^{i\omega_k t} + a_{-k} e^{-i\omega_k t}) e_k(x),$$

we have

$$E(0) \asymp \sum_{k=1}^{\infty} |\omega_k|^2 (|a_k|^2 + |a_{-k}|^2),$$

and

$$\int_{\Omega} |u'(t,x)|^2 \, dx = \sum_{k=1}^{\infty} \left| a_k i\omega_k e^{i\omega_k t} - a_{-k} i\omega_k e^{-i\omega_k t} \right|^2 \tag{8.18}$$

by the orthonormality of the functions e_k. For any given interval I we have

$$\int_I \left| a_k i\omega_k e^{i\omega_k t} - a_{-k} i\omega_k e^{-i\omega_k t} \right|^2 \, dt \asymp |\omega_k|^2 (|a_k|^2 + |a_{-k}|^2)$$

uniformly in k, because we may apply Ingham's Theorem 4.3 (p. 59) for all but finitely many indices k, and Theorem 4.6 (p. 67) for the remaining finitely many indices. (Since we have only two-term sums, these estimates can also be obtained by a direct computation, without applying these theorems.) Hence, integrating (8.18) over I, we obtain (8.17) with Ω in place of Ω_0.

On the other hand, (8.17) does not hold for an arbitrary (nonempty) open subset Ω_0 of Ω. Indeed, we have already seen in the analogous problem concerning a spherical plate in Section 7.5 (p. 143) that the domain of observation has to satisfy some geometric conditions.

Nevertheless, according to some theorems of Haraux [49] and Jaffard [54], [55], the rectangular plates enjoy a particular situation: the estimates (8.17) hold for *all* nonempty open subsets Ω_0 of Ω. Their result was extended to arbitrary dimensions in [65]; we have in fact the following result:

Proposition 8.8. *Let Ω be the product of N open intervals in $\mathbb{R}^N$, Ω_0 an arbitrary nonempty open subset of Ω, and I an arbitrary interval. Then the estimates (8.17) hold for all solutions of (8.16).*

In order to simplify the formulae, we prove this proposition here only in the special case

$$\ell_1 = \cdots = \ell_N = \pi \quad \text{and} \quad a_1 = \cdots = a_N = 0.$$

(In [65] it was assumed that $a_1 = \cdots = a_N = 0$.) At the end of the section we indicate the (small) changes to be done in the general case.

Beginning of the proof. The solutions of (8.16) are given by the formula

$$u(t,x) = \sum_{k \in \mathbb{N}^N} (a_k^+ e^{i|k|^2 t} + a_k^- e^{-i|k|^2 t}) \prod_{j=1}^{N} \sin k_j x_j$$

with suitable complex coefficients a_k^+ and a_k^-, where $\mathbb{N}$ denotes the set of (strictly) positive integers and we use the notation

$$k = (k_1, \ldots, k_N) \quad \text{and} \quad x = (x_1, \ldots, x_N).$$

Using Euler's formula

$$\sin z = \frac{e^{iz} - e^{-iz}}{2i},$$

we may rewrite it in the form

$$u(t, x) = \sum_{k \in (\mathbb{Z}^*)^N} b_k^+ e^{i(|k|^2 t + k \cdot x)} + b_k^- e^{-i(|k|^2 t + k \cdot x)}$$

with new complex coefficients b_k^+ and b_k^-, where $\mathbb{Z}^*$ denotes the set of nonzero integers.

Using this formula, a straightforward computation shows that

$$E(0) = \pi^N \sum_{k \in (\mathbb{Z}^*)^N} |k|^4 \left(|b_k^+|^2 + |b_k^-|^2 \right)$$

and

$$u'(t, x) = \sum_{k \in (\mathbb{Z}^*)^N} i|k|^2 b_k^+ e^{i(|k|^2 t + k \cdot x)} - i|k|^2 b_k^- e^{-i(|k|^2 t + k \cdot x)}.$$

Therefore, putting

$$c_k^+ := i|k|^2 b_k^+, \quad c_k^- := -i|k|^2 b_k^-, \quad \text{and} \quad \omega_k := (|k|^2, k) = (|k|^2, k_1, \ldots, k_N)$$

for brevity, the estimates (8.17) may be rewritten in the following form:

$$\int_{I \times \Omega_0} \left| \sum_{k \in (\mathbb{Z}^*)^N} c_k^+ e^{i\omega_k \cdot (t,x)} + c_k^- e^{i\omega_k \cdot (t,x)} \right|^2 dx \, dt \asymp \sum_{k \in (\mathbb{Z}^*)^N} \left(|c_k^+|^2 + |c_k^-|^2 \right).$$

Thus the proposition will be proved if we establish the following stronger result:

Lemma 8.9. Set $\omega_k := (|k|^2, k)$ for all $k \in \mathbb{Z}^N$. The following estimates hold for every nonempty, bounded, open set G in $\mathbb{R}^{N+1}$:

$$\int_G \left| \sum_{\substack{k \in \mathbb{Z}^N \\ k \neq 0}} c_k^+ e^{i\omega_k \cdot (t,x)} + c_k^- e^{i\omega_k \cdot (t,x)} \right|^2 dx \, dt \asymp \sum_{\substack{k \in \mathbb{Z}^N \\ k \neq 0}} \left(|c_k^+|^2 + |c_k^-|^2 \right). \tag{8.19}$$

It is sufficient to prove the estimates (8.19) for arbitrarily small balls $G = B_R$. Indeed, the general case then easily follows by covering G by a finite number of translates of B_R and by applying the triangle inequality, as we have already done several times before.

The following observation is crucial. Given a positive number R, assume that there exists a partition of the family

$$\{\pm\omega_k \; : \; k \in \mathbb{Z}^N, \quad k \neq 0\} \tag{8.20}$$

into finitely many, say m, subfamilies, each of which has a uniform gap γ_j such that

$$\frac{2\sqrt{\mu_1}}{\gamma_1} + \cdots + \frac{2\sqrt{\mu_1}}{\gamma_m} < R.$$

Here μ_1 denotes, as in Theorem 8.1 (p. 154), the first eigenvalue of $-\Delta$ in $H_0^1(B_1)$. Then (8.19) holds for $G = B_R$. Indeed, as a consequence of Theorem 8.1, the corresponding estimates hold for the subfamilies with $G = B_{R_j}$, $R_j = 2\sqrt{\mu_1}/\gamma_j$. Since $R > R_1 + \cdots + R_m$ and since the whole family also has a uniform gap (observe that all components of all vectors are integers), we conclude by applying Proposition 8.4 (p. 161).

The following result states that there exist such partitions for *every* $R > 0$; this will complete the proof of Proposition 8.8. In order to simplify the notation, let us denote by $d(A)$ the largest uniform gap of a set $A \subset \mathbb{R}^{N+1}$, i.e.,

$$d(A) := \inf\{\|a - a'\|_2 \; : \; a, a' \in A, \quad a \neq a'\}.$$

Note that $d(A) = \infty$ if A has at most one point.

Lemma 8.10. *For every given $\varepsilon > 0$, the family (8.20) has a finite partition $Y_1, \ldots, Y_m$ satisfying*

$$\frac{1}{d(Y_1)} + \cdots + \frac{1}{d(Y_m)} < \varepsilon.$$

Proof. Since the family (8.20) is the union of the congruent sets

$$Z := \{\omega_k \; : \; k \in \mathbb{Z}^N, \quad k \neq 0\} \quad \text{and} \quad -Z,$$

it suffices to prove that Z has the above-mentioned property.

Fix a small positive number δ, say[4] $0 < \delta < 1/(N+1)!$. A subset X of Z is called *r-thin* if there exist nonzero vectors $k_1, \ldots, k_r \in \mathbb{Z}^N$ and intervals $I_1, \ldots, I_r$ of length δ such that

$$-\delta \leq k_j \cdot k_n \leq \delta \quad \text{for all} \quad j \neq n$$

and

$$k_j \cdot k \in I_j \quad \text{for all} \quad k \in X \quad \text{and} \quad j = 1, \ldots, r.$$

Observe that Z itself is a 0-thin set. Furthermore, by the above choice of δ, an N-thin set has at most one element. Indeed, if an N-thin set had two distinct elements $(|m|^2, m)$ and $(|m + k_0|^2, m + k_0)$, then we would have for each j the relations

---

[4]This choice will be explained soon.

$$k_j \cdot (m + k_0) \in I_j \quad \text{and} \quad k_j \cdot m \in I_j,$$

implying

$$-\delta \le k_j \cdot k_0 \le \delta$$

because $|I_j| = \delta$. Since[5] $k_j \cdot k_j \ge 1$ for $j = 0, \ldots, N$, it would follow that the determinant

$$|k_j \cdot k_n|_{j,n=0}^{N}$$

was different from zero by our choice of δ. But this is impossible because the $N+1$ vectors $k_0, \ldots, k_N$ cannot be linearly independent in the N-dimensional space $\mathbb{R}^N$.

The following step is crucial. If X is an r-thin set and c a positive number, then let us denote by $X(c)$ the set of points $(|m|^2, m) \in X$ such that $|k \cdot m| \ge c^2$ for all nonzero vectors $k \in \mathbb{Z}^N$ satisfying

$$|k| < c, \quad \text{and} \quad |k_j \cdot k| \le \delta, \quad j = 1, \ldots, r. \tag{8.21}$$

We have

$$d(X(c)) \ge c \tag{8.22}$$

for all $c > 0$, i.e., $X(c)$ has a big uniform gap if we choose a sufficiently large c.

It suffices to show that if $(|m|^2, m)$ and $(|m + k|^2, m + k)$ are two distinct elements of $X(c)$, then at least one of the two following inequalities is satisfied:

$$|k| \ge c \quad \text{or} \quad \left| |m + k|^2 - |m|^2 \right| \ge c.$$

It follows again from the definition of X that $|k_j \cdot k| \le \delta$ for $j = 1, \ldots, r$. Therefore, if $|k| < c$, then we deduce from the definition of $X(c)$ that $|k \cdot m| \ge c^2$. Consequently, we have

$$\begin{aligned}
\left| |m + k|^2 - |m|^2 \right| &= |2m \cdot k - k \cdot k| \\
&\ge 2|m \cdot k| - |k|^2 \\
&> 2c^2 - c^2 \\
&= c^2.
\end{aligned}$$

The following step is decisive: *if X is an r-thin set for some $r < N$ and c is a positive number, then $X \backslash X(c)$ has a finite covering by $(r + 1)$-thin sets.* Indeed, let $\mathcal{I}$ be a finite set of intervals I of length δ, covering $[-c^2, c^2]$, and let $\mathcal{K}$ be the (finite) set of nonzero vectors $k \in \mathbb{Z}^N$ satisfying (8.21). Then, by the definition of $X(c)$ a suitable covering is given by the sets

$$\{(|m|^2, m) : k \cdot m \in I\}, \quad I \in \mathcal{I}, \quad k \in \mathcal{K}.$$

---

[5] We recall that all components are integers.

Now we are ready to complete the proof of the lemma. Since Z is a 0-thin set, choosing $c_1 > 1/\varepsilon$, setting $Y_1 := Z(c_1)$, and using the last observation, we obtain a finite covering

$$Z = Y_1 \cup X_2 \cup \cdots \cup X_p$$

of Z, where

$$\frac{1}{d(Y_1)} < \varepsilon$$

by (8.22), and $X_2, \ldots, X_p$ are 1-thin sets.

Next choose a sufficiently large positive number c_2 such that

$$\frac{1}{d(Y_1)} + \frac{p-1}{c_2} < \varepsilon.$$

Setting $Y_j := X_j(c_2)$ for $j = 2, \ldots, p$ and using the same observation as before, we obtain a finite covering

$$Z = Y_1 \cup Y_2 \cup \cdots \cup Y_p \cup X_{p+1} \cup \cdots \cup X_q$$

of Z such that

$$\frac{1}{d(Y_1)} + \cdots + \frac{1}{d(Y_p)} < \varepsilon$$

and $X_{p+1}, \ldots, X_q$ are 2-thin sets.

Continuing this construction, after N steps we obtain a finite covering

$$Z = Y_1 \cup Y_2 \cup \cdots \cup Y_r \cup X_{r+1} \cup \cdots \cup X_s$$

of Z such that

$$\frac{1}{d(Y_1)} + \cdots + \frac{1}{d(Y_r)} < \varepsilon$$

and $X_{r+1}, \ldots, X_s$ are N-thin sets. We conclude by recalling that $d(A) = \infty$ for all N-thin sets, so that

$$\frac{1}{d(Y_1)} + \cdots + \frac{1}{d(Y_r)} + \frac{1}{d(X_{r+1})} + \cdots + \frac{1}{d(X_s)} < \varepsilon.$$

The above proof can easily be adapted to the general case. First of all, if Ω is the product of intervals of differing lengths, then the above proof remains valid; we have only to modify the notation; see, e.g., [65], where the proof was carried out in the general case. Next, if lower-order terms are also allowed in the equations, then we may have a finite number of nonreal exponents. However, it follows from the asymptotic behavior of the eigenvalues that the influence of the lower-order terms is very small if we neglect a (sufficiently large) finite number of exponents w_k. In particular, the crucial Lemma 8.10 remains valid if we remove a finite number of exponents w_k. If none of the w_n's is equal to zero, then the theorem easily follows by applying Proposition 8.3

(p. 158) in the usual way. Otherwise, we have to generalize Proposition 8.3 in the direction outlined in a remark following the statement of this proposition. More precisely, denoting by N the set on indices k for which $\omega_k = 0$, we have to prove that if for some ball $B_0 \subset \mathbb{R}^{N+1}$ we have

$$\int_{B_0} \left| \sum_{k \in K \setminus N} x_k e^{\omega_k \cdot (t,x)} \right|^2 dt \asymp \sum_{k \in K \setminus N} |x_k|^2$$

for all *finite* sums with complex coefficients x_k, then for every ball B having a strictly bigger radius, we also have

$$\int_B \left| \sum_{k \in N} \beta_k t e^{\omega_k \cdot (t,x)} + \sum_{k \in K} x_k e^{\omega_k \cdot (t,x)} \right|^2 dt \, dt \asymp \sum_{k \in K} |x_k|^2$$

for all sums with square-summable complex coefficients β_k and x_k. This can be established by a simple adaptation of the proof of Theorem 6.2, which is left to the reader.

9

A General Ingham-Type Theorem

The main purpose of this section is to establish a far-reaching generalization of Ingham's Theorem 4.3 (p. 59). Our main Theorem 9.4, obtained in [8] and [9], will contain as a special case a celebrated improvement of Ingham's theorem, due to Beurling, who determined the critical length of the intervals for the validity of the estimates. Moreover, and this will be essential in our applications in the following chapter, we will also weaken the uniform gap condition. Our main result is motivated and formulated in Section 9.1, and its proof occupies the following four sections.

Ingham's original motivation was to give a simple proof of an important theorem of Pólya on singular points of certain Dirichlet series. Subsequently, Pólya's theorem was generalized by Bernstein. In the last Section 9.6 of this chapter we present a simple proof, given in [79], of Bernstein's theorem by following the ideas of Ingham but by applying our Theorem 9.4 instead of his Theorem 4.3.

The proof Theorem 9.4 is long, so the reader may prefer to read, just after Section 9.1, either Section 9.6 on the Dirichlet series, or the following Chapter 10, which is entirely based on this theorem. Having seen its usefulness, the reader can then return to learn the proof, too.

9.1 Generalization of a Theorem of Beurling

Let us begin improving Ingham's Theorem 4.3:

Theorem 9.1. *Assume that a sequence $(\lambda_n)_{n=-\infty}^{\infty}$ of real numbers satisfies for some $\gamma_1 > 0$ the gap condition*

$$\lambda_{n+1} - \lambda_n \geq \gamma_1 \quad \text{for all} \quad n. \tag{9.1}$$

Furthermore, assume that for some $\gamma > 0$ and an integer $M \geq 1$ we also have[1]

$$\lambda_{n+M} - \lambda_n \geq M\gamma \quad \text{for all} \quad n. \tag{9.2}$$

Then for any fixed interval I of length $|I| > 2\pi/\gamma$, all finite sums

$$f(t) = \sum_n b_n e^{i\lambda_n t}, \quad b_n \in \mathbb{C}, \tag{9.3}$$

satisfy the estimates

$$\int_I |f(t)|^2 \, dt \asymp \sum_n |b_n|^2. \tag{9.4}$$

Proof. Fix an interval J of length $|J| = |I|/M$. As a consequence of (9.2) we may apply Ingham's theorem to each of of the M subsequences $(\omega_{Mk+j})_{k=-\infty}^{\infty}$, $j = 1, \ldots, M$. As a consequence of hypothesis (9.1) we may conclude by applying Proposition 8.4 (p. 161).

Theorem 9.1 is optimal. In order to show this, let us introduce, following Pólya [113], the upper density D^+ of the sequence (λ_n). It is defined by the formula

$$D^+ := \lim_{r \to \infty} \frac{n^+(r)}{r},$$

where $n^+(r)$ denotes the largest number of terms of the sequence (λ_n) contained in an interval of length r. The existence of the limit follows from the subadditivity relation $n^+(r + s) \leq n^+(r) + n^+(s)$ for all $r, s > 0$. Indeed, set $\alpha = \inf_{r>0} n^+(r)/r$, choose $\beta > \alpha$ arbitrarily, and then choose $R > 0$ such that $n^+(R)/R < \beta$. For every $r > R$, writing $r = mR + s$ with an integer $m \geq 1$ and with $0 \leq s < R$, we have

$$\alpha \leq \frac{n^+(r)}{r} \leq \frac{n^+((m+1)R)}{mR} \leq \frac{(m+1)n^+(R)}{mR} = \frac{m+1}{m} \frac{n^+(R)}{R}.$$

If $r \to \infty$, then $m \to \infty$, and therefore $n^+(r)/r < \beta$ for all sufficiently large r. Letting $\beta \to \alpha$, we conclude that $n^+(r)/r \to \alpha$ as $r \to \infty$.

According to a celebrated theorem of Beurling [14] (see also Seip [124]), the critical length is equal to $2\pi D^+$:

Theorem 9.2. *(Beurling) Let (λ_n) satisfy the condition (9.1) again. Then the estimates (9.4) hold if $|I| > 2\pi D^+$, and they do not hold if $|I| < 2\pi D^+$.*

A counterexample of Ingham [52] shows that the estimates (9.4) can fail in the limiting case $|I| = 2\pi D^+$.

---

[1]Of course, this condition is satisfied with $\gamma = \gamma_1$ and $M = 1$. The point is that we can sometimes choose a number $\gamma > \gamma_1$ that leads to a weakening of the condition on the length of I in the estimate (9.4).

Example. Consider the sequence (λ_n) given by

$$\lambda_{2n} = 2n \quad \text{and} \quad \lambda_{2n+1} = 2n + 10^{-3}$$

for all n. Then condition (9.1) is satisfied with $\gamma_1 = 10^{-3}$, so that Ingham's theorem implies the estimates (9.4) for every bounded interval I of length $> 2000\pi$.

On the other hand, we have $D^+ = 1$ here, so that by Beurling's theorem the estimates (9.4) hold in fact for every bounded interval I of length $> 2\pi$, and they do not hold if I is shorter than 2π.

Example. Consider the sequence (λ_n) given by $\lambda_n = n^3$ for all n. Then condition (9.1) is satisfied with $\gamma_1 = 1$, so that Ingham's theorem implies the estimates (9.4) for every bounded interval I of length $> 2\pi$.

On the other hand, now we have $D^+ = 0$, so that by Beurling's theorem the estimates (9.4) hold in fact for *every* interval I.

Let us show that the assumptions of the preceding two theorems on the length of I are equivalent; hence the sufficient part of Beurling's theorem is equivelent to Theorem 9.1:

Proposition 9.3. *Given an increasing sequence (λ_n) and an interval I, the condition $|I| > 2\pi D^+$ is equivalent to the existence of a number $\gamma > 0$ and of an integer $M \geq 1$ such that $|I| > 2\pi/\gamma$ and*

$$\lambda_{n+M} - \lambda_n \geq M\gamma \quad \text{for all} \quad n. \tag{8.2}$$

Proof. Condition (9.2) implies at once that, more generally,

$$\lambda_{n+kM} - \lambda_n \geq kM\gamma \quad \text{for all} \quad n$$

for $k = 1, 2, \ldots$. These relations imply the inequalities

$$n^+(kM\gamma) \leq kM + 1, \quad k = 1, 2, \ldots.$$

Letting $k \to \infty$, we conclude that $D^+ \leq \gamma^{-1}$, and therefore the condition $|I| > 2\pi/\gamma$ implies that $|I| > 2\pi D^+$.

Conversely, given a bounded interval of length $|I| > 2\pi D^+$, choose a number $\gamma > 0$ such that

$$|I| > \frac{2\pi}{\gamma} > 2\pi D^+,$$

and then choose a sufficiently large r such that

$$\frac{2\pi}{\gamma} > 2\pi \frac{n^+(r)}{r}.$$

We may assume that $M := r/\gamma$ is an integer. Then we have $n^+(r) < M$, which implies that

$$\lambda_{n+M} - \lambda_n > r \quad \text{for all} \quad n.$$

Since $r = M\gamma$, this is just the condition (9.2).

The preceding theorems do not apply to the following example:

Example. Consider the increasing sequence (λ_n) of the numbers n^3 and $n^3 + n^{-3}$, where n runs over all odd integers. Then condition (9.2) is satisfied with $M = 2$ and $\gamma = 1.5$, but we do not have a uniform gap γ_1. Applying Theorem 3 in [6] (proved in Theorem 2.1 in [7]), an improvement of earlier results of Jaffard, Tucsnak, and Zuazua [56], [57] and of Castro and Zuazua [23], we obtain for every bounded interval I of length $> 3\pi$ the estimate

$$\int_I |f(t)|^2 \, dt \asymp \sum |b_n + b_{n-1}|^2 + |n|^{-3}\big(|b_n|^2 + |b_{n-1}|^2\big),$$

where n runs over the odd integers.

In view of second example above it is natural to expect that the estimates of the third example hold in fact for *every* bounded interval I. We will confirm this conjecture by applying a generalization of Theorem 9.1 for sequences not necessarily having a uniform gap. In order to formulate our main result, we introduce the divided differences of exponential functions with close exponents.

Let $(\lambda_n)_{n=-\infty}^{\infty}$ be an increasing sequence of real numbers, satisfying for some integer $M \geq 1$ and a real number $\gamma > 0$ the condition (9.2). (For $M = 1$ this assumption implies that the sequence is *strictly* increasing. On the other hand, for $M > 1$ we can have repeated terms, but no term can be repeated more than M times.)

Fix a number $0 < \gamma' \leq \gamma$ and denote by A_j $(j = 1, \ldots, M)$ the set of integers m satisfying the following conditions:

$$\begin{cases} \lambda_m - \lambda_{m-1} \geq \gamma', \\ \lambda_n - \lambda_{n-1} < \gamma' \quad \text{for} \quad m+1 \leq n \leq m+j-1, \\ \lambda_{m+j} - \lambda_{m+j-1} \geq \gamma'. \end{cases}$$

We say sometimes that the exponents $\lambda_m, \ldots, \lambda_{m+j-1}$ form a *chain of close exponents* for γ'. Then the $M(M+1)/2$ sets

$$A_j + k := \{n + k \; : \; n \in A_j\}, \quad 0 \leq k < j \leq M,$$

are disjoint. Furthermore, since by (9.2) we cannot have M consecutive distances $\lambda_{n+1} - \lambda_n < \gamma'$, they form a partition of the set $\mathbb{Z}$ of *all* integers.

Let us introduce for $m \in A_j$ the divided differences $e_m(t), \ldots, e_{m+j-1}(t)$ of the exponential functions

$$f_n(t) := e^{i\lambda_n t}, \quad n = m, \ldots, m+j-1,$$

by the formula (see, e.g., [53], theorem 2, p. 250)

$$e_n(t) := (it)^{n-m} \int_0^1 \int_0^{s_1} \cdots \int_0^{s_{n-m-1}} A \, ds_{n-m} \cdots ds_1 \qquad (9.5)$$

with

$$A := \exp\Big(i(s_{n-m}[\lambda_n - \lambda_{n-1}] + \cdots + s_1[\lambda_{m+1} - \lambda_m] + \lambda_m)t\Big)$$

for $n = m, \ldots, m + j - 1$. We have $e_m = f_m$ in particular.

If $\lambda_m, \ldots, \lambda_n$ are distinct, then (9.5) is equivalent to the familiar expression

$$e_n(t) := \sum_{p=m}^{n} \Big[\prod_{q=m}^{n}{}'(\lambda_p - \lambda_q)\Big]^{-1} f_p(t),$$

where the sign $'$ in the products indicates the omission of the zero factor corresponding to $q = p$.

In the other extreme case, in which $\lambda_m = \cdots = \lambda_n$, we have $e_n(t) = t^{n-m} e^{i\lambda_n t}$.

Let us consider the (finite) sums of the form

$$f(t) = \sum_{n=-\infty}^{\infty} a_n e_n(t), \quad a_n \in \mathbb{C}. \tag{9.6}$$

If the sequence (λ_n) is strictly increasing, then the set of functions of the forms (9.3) and (9.6) coincide. In the general case (see Lemma 9.5 below) the set of functions (9.6) is the linear span of the functions $t^j \exp(i\lambda_n t)$, where n runs over the integers and $j = 0, \ldots, k - 1$ if λ_n appears k times in the sequence (λ_n). In particular, the set of functions of the form (9.6) always contains those of the form (9.3).

Theorem 9.4. *Given a (not necessarily strictly) increasing sequence (λ_n) of real numbers satisfying*

$$\lambda_{n+M} - \lambda_n \geq M\gamma \quad \text{for all} \quad n, \tag{8.2}$$

fix $0 < \gamma' \leq \gamma$ arbitrarily and introduce the sequence of functions (e_n) as described above. The sums (9.6) satisfy the estimates

$$\int_I |f(t)|^2 \, dt \asymp \sum_{n=-\infty}^{\infty} |a_n|^2 \tag{9.7}$$

for every interval I of length $|I| > 2\pi\gamma$. The corresponding constants c_1 and c_2 depend only on M, γ, γ', and on the interval I.

Moreover, for every real number $\sigma > M - (1/2)$ we also have the estimates

$$\int_{-\infty}^{\infty} (1 + t^2)^{-\sigma} |f(t)|^2 \, dt \asymp \sum_{n=-\infty}^{\infty} |a_n|^2 \tag{9.8}$$

with corresponding constants depending on M, γ, γ', and σ.

Several remarks are in order.

Remarks.

- According to Proposition 9.3, the estimates (9.7) hold (for a suitable choice of the functions e_n) whenever $|I| > 2\pi D^+$. Mehrenberger established recently in [108] that no estimate of this type can hold if $|I| < 2\pi D^+$.
- The liberty of choosing $0 < \gamma' \leq \gamma$ instead of $\gamma' = \gamma$ is useful because the structure of the sequence (e_n) may become simpler for a smaller γ'. Indeed, by diminishing γ' some chains of close exponents for γ can break into several shorter chains of close exponents for γ'. For example, if the sequence (λ_n) satisfies not only (9.2) but also the gap condition (9.1) with some γ_1 (which may be much smaller than γ), then choosing $0 < \gamma' \leq \min(\gamma_1, \gamma)$, we have $A_1 = \mathbb{Z}$ and $A_2 = \cdots = A_M = \emptyset$, so that Theorem 9.4 reduces to Beurling's theorem.
- Consider an extreme case in which the sequence (λ_n) is not *strictly* increasing. If every term of the sequence (λ_n) is repeated M times, then we obtain that the estimates (9.4) hold for all sums of the form

$$f(t) = \sum_{k=-\infty}^{\infty} \sum_{p=0}^{M-1} a_{Mk+p} t^p e^{i\lambda_{Mk} t}, \quad a_n \in \mathbb{C}.$$

This was proved in a different way in [7] (where the case of several variables was also studied). If the exponents are the integers, then this result was proved earlier by Ullrich [133] (in that special case he also established the estimates for intervals of critical length).

Example. Applying Theorem 9.4, we obtain that the estimates of the third example hold for every (arbitrarily small) interval I.

We prove Theorem 9.4 in Sections 9.2–9.4 by generalizing a method of Kahane [61]. Part of his approach is simplified by applying the method of Haraux [48].

In this chapter the letters c, c_1, c_2, etc., will denote various (strictly) positive constants depending only on M, γ, γ', and on the intervals I and/or J under consideration.

9.2 Chains of Close Exponents

In the rest of this chapter we assume that the assumptions of Theorem 9.4 are fulfilled. In order to simplify the notation we shall write $p \sim q$ if there exists $m \in A_j$ such that both p and q belong to the set $\{m, \ldots, m+j-1\}$, i.e., if λ_p and λ_q belong to the same chain of close exponents beginning with λ_m and ending with λ_{m+j-1}.

Our first lemma shows that if Theorem 9.4 is established for some particular value of $0 < \gamma' \leq \gamma$, then it holds for *all* values $0 < \gamma' \leq \gamma$. This will be useful later, because the proof of Theorem 9.4 will be simplified by assuming that γ' is sufficiently small.

Let

$$\lambda_m \leq \cdots \leq \lambda_{m+j-1}$$

be a chain of close exponents for $\gamma' = \gamma$. Fix a number $0 < \gamma' < \gamma$, then this chain breaks into r chains of close exponents for γ', say

$$\lambda_m \leq \cdots \leq \lambda_{m+j_1-1}, \ldots, \lambda_{m+j_{r-1}} \leq \cdots \leq \lambda_{m+j_r-1}, \quad j_r = j.$$

(We do not exclude the case $r = 1$ when the chain does not break, but then the following lemma is obvious.)

Lemma 9.5. *Let us denote by (E_n) and (e_n), respectively, the sequences of divided differences in Theorem 9.4 corresponding to the choices $\gamma' = \gamma$ and $0 < \gamma' < \gamma$ as above.*

Then both sequences

$$e_m, \ldots, e_{m+j-1} \quad and \quad E_m, \ldots, E_{m+j-1}$$

are linearly independent, and they span the same vector space.

Moreover, there exist two positive constants $c_1, c_2 > 0$, depending only on M, γ, and γ' such that

$$c_1 \sum_{n \sim m} |a_n|^2 \leq \sum_{n \sim m} |A_n|^2 \leq c_2 \sum_{n \sim m} |a_n|^2$$

whenever

$$\sum_{n \sim m} a_n e_n = \sum_{n \sim m} A_n E_n.$$

Proof. Assume, for example, that the chain

$$\lambda_m \leq \cdots \leq \lambda_{m+j-1}$$

of close exponents for γ breaks into r chains of close exponents for γ' as indicated above. It follows directly from the definitions that

$$\gamma' \leq \lambda_p - \lambda_q < M\gamma \tag{9.9}$$

whenever $m \leq q < m + j_k$ and $m + j_k \leq p < m + j$ for some k, and that

$$e_n = E_n \quad \text{if} \quad m \leq n < m + j_1. \tag{9.10}$$

Furthermore, we claim that

$$e_{m+j_k} = \sum_{p=m}^{m+j_k} \prod_{q=m}^{p-1} (\lambda_{m+j_k} - \lambda_q) E_p \tag{9.11}$$

and

$$e_p = \prod_{q=m}^{m+j_k-1} (\lambda_p - \lambda_q) E_p \quad \text{if} \quad m + j_k < p < m + j_{k+1}, \tag{9.12}$$

for $k = 1, \ldots, r - 1$.

Indeed, we deduce from Newton's interpolational formula

$$f_n = \sum_{p=m}^{n} \prod_{q=m}^{p-1} (\lambda_n - \lambda_q) E_p, \quad n = m, \ldots, m + j - 1,$$

that

$$f_n = f_{m+j_k} + \sum_{p=m+j_k+1}^{n} \prod_{q=m}^{p-1} (\lambda_n - \lambda_q) E_p \quad \text{if} \quad m + j_k \le p < m + j_{k+1}. \tag{9.13}$$

Furthermore, applying Newton's interpolational formula for the sequence (e_n), we also have

$$f_n = e_{m+j_k} + \sum_{p=m+j_k+1}^{n} \prod_{q=m+j_k}^{p-1} (\lambda_n - \lambda_q) e_p \quad \text{if} \quad m + j_k \le p < m + j_{k+1}. \tag{9.14}$$

Since $e_{m+j_k} = f_{m+j_k}$, comparing (9.13) and (9.14) shows that the relations (9.11) and (9.12) follow.

Let us show that the sequence $E_m, \ldots, E_{m+j-1}$ is linearly independent. Assume that the exponents $\lambda_m, \ldots, \lambda_{m+j-1}$ take r distinct values $\lambda_{m_1}, \ldots, \lambda_{m_r}$, with multiplicities $j_1, \ldots, j_r$, respectively, so that $j_1 + \cdots + j_r = j$. Then, choosing a sufficiently small γ', we obtain that the functions $e_m, \ldots, e_{m+j-1}$ are constant multiples of $t^q e^{i\lambda_{m_s} t}$, $s = 1, \ldots, r$, $q = 0, \ldots, j_r - 1$. It is well known from the theory of ordinary linear differential equations (see, e.g., [26]) that the latter system is linearly independent.

It follows from (9.9)–(9.12) that $e_m, \ldots, e_{m+j-1}$ all belong to the linear span of $E_m, \ldots, E_{m+j-1}$. Hence the latter is at least j-dimensional. This proves the linear independence of $E_m, \ldots, E_{m+j-1}$.

Now let us turn back to the original choice of γ'. It follows from (9.9)–(9.12) that

$$e_n = \sum_{p=m}^{n} \alpha_{np} E_p, \quad n = m, \ldots, m + j - 1, \tag{9.15}$$

with suitable coefficients α_{np} satisfying

$$\min\{1, \gamma'\}^M \le |\alpha_{np}| \le \max\{1, M\gamma\}^M. \tag{9.16}$$

Hence

$$\sum_{n \sim m} a_n e_n = \sum_{p=m}^{n} \sum_{n \sim m} a_n \alpha_{np} E_p$$

for all $n \sim m$, so that

$$|A_p| = \left| \sum_{n \sim p} a_n \alpha_{np} \right| \le c \max\{|a_m|, \ldots, |a_p|\}$$

if $p \sim m$, and therefore

$$\sum_{p \sim m} |A_p|^2 \le c_2 \sum_{p \sim m} |a_p|^2,$$

with some constants c, c_2 depending only on M and γ.

Now observe that the transformation matrix in (9.15) is triangular. Therefore, using (9.16) we obtain the relations

$$E_p = \sum_{n=m}^{p} \beta_{pn} e_n, \quad p = m, \ldots, m+j-1, \tag{9.17}$$

with suitable coefficients β_{pn}, bounded by some constant depending only on M, γ, and γ'. Since the system $E_m, \ldots, E_{m+j-1}$ is already known to be linearly independent, we deduce from these relations that $e_m, \ldots, e_{m+j-1}$ are also linearly independent.

Finally, it follows from the relations (9.17) as above that

$$|a_n| \le c \max\{|A_m|, \ldots, |A_n|\}, \quad n \sim m,$$

and therefore

$$\sum_{n \sim m} |a_n|^2 \le c_1 \sum_{n \sim m} |A_n|^2,$$

with some constants c, c_1 depending only on M, γ, and γ'.

In the sequel the sequence (e_n) will correspond to the fixed value of γ' as given in Theorem 9.4.

First we establish the *direct inequality* for each chain of close exponents, by giving an estimate of the constants as functions of the integral limits. This will be needed later.

Lemma 9.6. *Fix a number $R > 0$. There exists a constant $c_1 = c_1(R)$ such that for every $m \in A_j$ $(1 \le j \le M)$ and for every real number b, all linear combinations of $e_m, \ldots, e_{m+j-1}$ satisfy the estimates*

$$\int_{b-R}^{b+R} \left| \sum_{p \sim m} a_p e_p \right|^2 dt \le c_1 (1 + b^2)^{M-1} \sum_{p \sim m} |a_p|^2.$$

Proof. First we deduce from the formula (9.5) the inequalities $|e_p(t)| \le |t|^{p-m}$ for all real t and for all $m \sim p$. Then we have

$$\int_{b-R}^{b+R} \left| \sum_{p \sim m} a_p e_p \right|^2 dt \leq j \sum_{p \sim m} \int_{b-R}^{b+R} |a_p e_p|^2 \, dt$$

$$\leq j \sum_{p \sim m} \int_{b-R}^{b+R} |t|^{2p-2m} \, dt \cdot |a_p|^2$$

$$\leq 2Rj \sum_{p \sim m} (R + |b|)^{2(p-m)} \cdot |a_p|^2$$

$$\leq 2Rj^2 (1 + R + |b|)^{2j-2} |a_p|^2.$$

Since $j \leq M$, the lemma follows.

Finally, we establish the *inverse inequality* for each chain of close exponents.

Lemma 9.7. *For every interval I there exists a constant $c > 0$ such that for every $m \in A_j$ ($1 \leq j \leq M$), all linear combinations of $e_m, \ldots, e_{m+j-1}$ satisfy the estimates*

$$\sum_{p \sim m} |a_p|^2 \leq c \int_I \left| \sum_{p \sim m} a_p e_p \right|^2 dt. \tag{9.18}$$

Proof. It follows from the linear independence of $e_m, \ldots, e_{m+j-1}$ that the estimates (9.18) are satisfied for every choice of $\lambda_m, \ldots, \lambda_{m+j-1}$. The proof will be completed if we show that we can choose the same constant $c > 0$ in (9.18) for each fixed $1 \leq j \leq M$, for all choices of $\lambda_m \leq \cdots \leq \lambda_{m+j-1}$ satisfying $\lambda_{m+j-1} - \lambda_m \leq M\gamma$. For each choice of these exponents, let us denote by $C(\lambda_m, \ldots, \lambda_{m+j-1})$ the smallest number c satisfying (9.18). This function is continuous because the formula (9.5) shows that the functions e_p depend continuously on numbers $\lambda_m, \ldots, \lambda_{m+j-1}$.

Observe that the integral in (9.18) does not change if we add the same real number to each λ_p. We may therefore restrict ourselves to the choices of numbers

$$0 = \lambda_m \leq \cdots \leq \lambda_{m+j-1} \leq M\gamma.$$

Since this is a compact set, the continuous function $C(\lambda_m, \ldots, \lambda_{m+j-1})$ has a finite maximum c on it. The proof is complete.

9.3 Proof of the Direct Part of Theorem 9.4

The following two lemmas establish the easier half of Theorem 9.4.

Lemma 9.8. *For every interval I, all sums*

$$f(t) = \sum_{n=-\infty}^{\infty} a_n e_n(t) \tag{9.19}$$

satisfy the estimate

$$\int_I |f(t)|^2 \, dt \le c \sum_{n=-\infty}^{\infty} |a_n|^2$$

with a constant c depending only on M, γ, γ', and on the interval I.

Proof. We may assume by a scaling argument that $\gamma' = M\pi$. Consider the functions H and h introduced at the beginning of Section 4.2, p. 62. Set $K := H * \cdots * H$ with $2M$ factors in the convolution, and let us denote by $k = h^{2M}$ its Fourier transform. Then K, k are continuous functions on the real line, and we have the following properties:

$$K(x) = 0 \quad \text{if} \quad |x| \ge M\pi,$$
$$k(0) = 4^M > 1,$$
$$0 \le k(t) \le C(1 + t^2)^{-2M} \quad \text{for all real} \quad t,$$

with some constant C. Let us fix a number $r > 0$ such that $k \ge 1$ in $[-r, r]$. It follows that for every fixed real number a there exists a constant c_2 such that

$$0 \le k(t - a) \le c_2(1 + b^2)^{-2M} \quad \text{whenever} \quad b - 1 < t < b + 1. \qquad (9.20)$$

Let us denote by K_a the inverse Fourier transform of $k(t-a)$ given by $K_a(x) = e^{-iax} K(x)$.

We have the following estimate:

$$\int_{a-r}^{a+r} |f(t)|^2 \, dt \le \int_{-\infty}^{\infty} k(t - a)|f(t)|^2 \, dt$$

$$= 2\pi \sum_{j=1}^{M} \sum_{m \in A_j} \sum_{p,q \sim m} K_a(\lambda_p - \lambda_q) a_p \overline{a_q}$$

because $K_a(\lambda_p - \lambda_q) = 0$ whenever $|\lambda_p - \lambda_q| \ge M\pi$. Hence

$$\int_{a-r}^{a+r} |f(t)|^2 \, dt \le \sum_{j=1}^{M} \sum_{m \in A_j} \int_{-\infty}^{\infty} k(t - a) \left| \sum_{p \sim m} a_p e_p(t) \right|^2 dt. \qquad (9.21)$$

Applying Lemma 9.6 with $R = 1/2$, $b \in \mathbb{Z}$, and using (9.20), we see that every integral on the right-hand side of (9.21) can be majorized as follows:

$$\int_{-\infty}^{\infty} k(t - a) \left| \sum_{p \sim m} a_p e_p(t) \right|^2 dt$$

$$\le \sum_{b=-\infty}^{\infty} c_1 c_2 (1 + b^2)^{-M-1} \sum_{p \sim m} |a_p|^2 =: c \sum_{p \sim m} |a_p|^2.$$

Therefore, we deduce from (9.21) the estimate

$$\int_{a-r}^{a+r} |f(t)|^2 \, dt \le c \sum_{j=1}^{M} \sum_{m \in A_j} \sum_{p \sim m} |a_p|^2 = c \sum_{n=-\infty}^{\infty} |a_n|^2.$$

Now, every interval I can be covered by finitely many intervals of length $2r$, say by $I_1, \ldots, I_m$. Then we have

$$\int_I |f(t)|^2 \, dt \le \sum_{p=1}^{m} \int_{I_p} |f(t)|^2 \, dt \le (c_1 + \cdots + c_m) \sum_{n=-\infty}^{\infty} |a_n|^2,$$

and the lemma follows.

Lemma 9.9. *For every real number $\sigma > M - (1/2)$, all sums (9.19) satisfy the estimates*

$$\int_{-\infty}^{\infty} (1 + t^2)^{-\sigma} |f(t)|^2 \, dt \le c \sum_{n=-\infty}^{\infty} |a_n|^2$$

with a constant c depending only on M, γ, γ', and σ.

Proof. Fix r as in the preceding lemma. Fix a large positive integer M' (to be chosen later) and introduce a function k as in the proof of Lemma 9.8, but by defining K as the convolution product of M' factors H instead of $2M$ factors. Then we have the following chain of inequalities (the constants c depend on r, and at the end we apply Lemma 9.6 in order to introduce the factor $(1 + b^2)^{M-1}$):

$$\int_{-\infty}^{\infty} (1 + t^2)^{-\sigma} |f(t)|^2 \, dt$$

$$= \sum_{a=-\infty}^{\infty} \int_{ar}^{(a+1)r} (1 + t^2)^{-\sigma} |f(t)|^2 \, dt$$

$$\le c \sum_{a=-\infty}^{\infty} (1 + a^2)^{-\sigma} \int_{ar}^{(a+1)r} |f(t)|^2 \, dt$$

$$\le c \sum_{a=-\infty}^{\infty} (1 + a^2)^{-\sigma} \int_{-\infty}^{\infty} k(t - ar) |f(t)|^2 \, dt$$

$$= c \sum_{j=1}^{M} \sum_{m \in A_j} \Big(\sum_{a=-\infty}^{\infty} (1 + a^2)^{-\sigma} \int_{-\infty}^{\infty} k(t - ar) \Big| \sum_{p \sim m} a_p e_p(t) \Big|^2 \, dt \Big)$$

$$= c \sum_{j=1}^{M} \sum_{m \in A_j} \Big(\sum_{a=-\infty}^{\infty} (1 + a^2)^{-\sigma} \sum_{b=-\infty}^{\infty} \int_{br}^{(b+1)r} k(t - ar) \Big| \sum_{p \sim m} a_p e_p(t) \Big|^2 \, dt \Big)$$

$$\leq c \sum_{j=1}^{M} \sum_{m \in A_j} \left(\sum_{a=-\infty}^{\infty} (1+a^2)^{-\sigma} \sum_{b=-\infty}^{\infty} (1+(b-a)^2)^{-M'} \right.$$

$$\left. \times \int_{br}^{(b+1)r} \left| \sum_{p \sim m} a_p e_p(t) \right|^2 dt \right)$$

$$\leq c \left(\sum_{j=1}^{M} \sum_{m \in A_j} \sum_{p \sim m} |a_p|^2 \right) \left(\sum_{a=-\infty}^{\infty} (1+a^2)^{-\sigma} \right.$$

$$\left. \times \sum_{b=-\infty}^{\infty} (1+(b-a)^2)^{-M'} (1+b^2)^{M-1} \right)$$

$$= cK \sum_{n=-\infty}^{\infty} |a_n|^2$$

with

$$K := \sum_{a=-\infty}^{\infty} (1+a^2)^{-\sigma} \sum_{b=-\infty}^{\infty} (1+(b-a)^2)^{-M'} (1+b^2)^{M-1}.$$

It remains to show that $K < \infty$. Choose a (small) $\varepsilon > 0$ such that

$$-\sigma + M - 1 + \varepsilon < -\frac{1}{2};$$

this is possible by our assumption on σ. Assume for the moment that

$$\sum_{b=-\infty}^{\infty} (1+(b-a)^2)^{-M'} (1+b^2)^{M-1} \leq c(1+a^2)^{M-1+\varepsilon} \qquad (9.22)$$

with a constant c that does not depend on a. Then

$$K \leq c \sum_{a=-\infty}^{\infty} (1+a^2)^{-\sigma+M-1+\varepsilon};$$

since the last series converges, we conclude that $K < \infty$.

For the proof of (9.22) assume by symmetry that $a \geq 0$. Choosing $M' \geq M$, we have

$$\sum_{b=a+1}^{\infty} (1+(b-a)^2)^{-M'} (1+b^2)^{M-1} \qquad (9.23)$$

$$= \sum_{n=1}^{\infty} (1+n^2)^{-M'} (1+(a+n)^2)^{M-1}$$

$$\leq 4 \sum_{n=1}^{\infty} n^{-2M'} (1+a^2)^{M-1} n^{2M-2}$$

$$\leq 7(1+a^2)^{M-1}$$

and

$$\sum_{b=-\infty}^{-1} \left(1 + (b-a)^2\right)^{-M'} (1+b^2)^{M-1} \tag{9.24}$$

$$= \sum_{n=1}^{\infty} (1 + (a+n)^2)^{-M'} (1+n^2)^{M-1}$$

$$\leq \sum_{n=1}^{\infty} (1+n^2)^{M-1-M'}$$

$$\leq \sum_{n=1}^{\infty} (1+n^2)^{-1} < \infty.$$

Now choose $q > 1$ such that $1/q < \varepsilon$, then $p > 1$ such that $p^{-1} + q^{-1} = 1$, and finally $M' \geq M$ such that $2M'p > 1$. Using the Hölder inequality we have

$$\sum_{b=0}^{a} \left(1 + (b-a)^2\right)^{-M'} (1+b^2)^{M-1}$$

$$\leq \left(\sum_{b=0}^{a} \left(1 + (b-a)^2\right)^{-M'p}\right)^{1/p} \left(\sum_{b=0}^{a} (1+b^2)^{(M-1)q}\right)^{1/q}$$

$$\leq c \left(\sum_{b=0}^{a} (1+b^2)^{(M-1)q}\right)^{1/q}.$$

Since

$$\left(\sum_{b=0}^{a} (1+b^2)^{(M-1)q}\right)^{1/q} \leq (1+a)^{1/q} (1+a^2)^{M-1}$$

$$\leq (1+a^2)^{M-1+(1/q)}$$

$$\leq (1+a^2)^{M-1+\varepsilon},$$

we conclude that

$$\sum_{b=0}^{a} \left(1 + (b-a)^2\right)^{-M'} (1+b^2)^{M-1} \leq c(1+a^2)^{M-1+\varepsilon}. \tag{9.25}$$

The inequality (9.22) now follows from (9.23), (9.24), and (9.25).

9.4 Biorthogonal Sequences

Now we turn to the proof of the inverse inequality. We recall the notation $f_n(t) := e^{i\lambda_n t}$ for the exponential functions.

We shall use the congruence notation modulo M: we write $n \equiv k$ if M divides $n - k$, i.e., if $n \in k + M\mathbb{Z}$. Furthermore, every function $\varphi \in L^2(I)$, where I is an interval, will be automatically extended by zero to the whole real line, so that $\varphi \in L^2(\mathbb{R})$ and $\varphi = 0$ in $\mathbb{R}\backslash I$.

Lemma 9.10. *Fix an integer* $1 \le k \le M$. *For every interval* I *of length* $|I| > 2\pi/(M\gamma)$ *we have*

$$\int_I \left| \sum_{n \equiv k} b_n f_n \right|^2 dt \asymp \sum_{n \equiv k} |b_n|^2$$

for all finite sums with complex coefficients b_n. *(The corresponding constants depend only on* M, γ, *and* $|I|$.*)*

Proof. As a consequence of condition (9.2) we have

$$\lambda_m - \lambda_n \ge M\gamma$$

whenever $m \equiv n$ and $m > n$. We conclude by applying Ingham's Theorem 9.1.

Lemma 9.11. *Fix an integer* $1 \le k \le M$ *and an integer* $m \in A_j$ *for some* $1 \le j \le M$. *For every interval* J *of length* $|J| > 2\pi/(M\gamma)$ *we have*

$$\int_J \left| \sum_{n \sim m} a_n e_n + \sum_{\substack{n \equiv k \\ n \not\sim m}} b_n f_n \right|^2 dt \asymp \sum_{n \sim m} |a_n|^2 + \sum_{\substack{n \equiv k \\ n \not\sim m}} |b_n|^2$$

for all finite sums with complex coefficients a_n *and* b_n. *(The corresponding constants depend only on* M, γ, γ', *and* $|J|$.*)*

This result is stronger than the preceding lemma because we added all close exponents belonging to the chain of λ_m.

Proof. The direct inequality follows from Lemmas 9.7 and 9.10 by applying the triangle inequality. Indeed, writing

$$f := \sum_{n \sim m} a_n e_n + \sum_{\substack{n \equiv k \\ n \not\sim m}} b_n f_n =: g + h \tag{9.26}$$

for brevity, we have

$$\int_J |f|^2 \, dt \le \int_J 2|g|^2 + 2|h|^2 \, dt \le c\left(\sum_{n \sim m} |a_n|^2 + \sum_{\substack{n \equiv k \\ n \not\sim m}} |b_n|^2 \right)$$

with a constant depending only on M, γ, γ', and $|J|$.

For the proof of the inverse inequality fix an interval $I = (a, b)$ of length $|I| > 2\pi/(M\gamma)$ and a number $r > 0$ such that $J = (a - jr, b + jr)$. Similarly to the proof of Theorem 6.2 (p. 93), let us introduce the product

$$I := I_{\lambda_m} \cdots I_{\lambda_{m+j-1}}$$

of the operators

$$(I_{\lambda_k} f)(t) := f(t) - \frac{1}{2r} \int_{-r}^{r} e^{-i\lambda_k s} f(t + s) \, ds, \quad k = m, \ldots, m + j - 1.$$

Applying the scalar case of Lemma 6.3 (p. 96) j times, we obtain that for f given by (9.26), If has the form

$$(If)(t) = \sum_{n \equiv k, \; n \not\sim m} b'_n f_n,$$

and that the following inequalities are satisfied with a constant c depending only on M and γ':

$$\int_I |(If)(t)|^2 \, dt \le c \int_J |f(t)|^2 \, dt,$$

$$|b_n| \le c|b'_n| \quad \text{for all} \quad n.$$

Applying Lemma 9.10 for If, we conclude that

$$\sum_{n \equiv k, \; n \not\sim m} |b_n|^2 \le c \int_J |f(t)|^2 \, dt, \tag{9.27}$$

and then also that

$$\int_J |h(t)|^2 \, dt \le c \int_J |f(t)|^2 \, dt.$$

Hence

$$\int_J |g(t)|^2 \, dt \le \int_J 2|f(t)|^2 + 2|h(t)|^2 \, dt \le c \int_J |f(t)|^2 \, dt.$$

Applying Lemma 9.7, we see that the last inequality implies that

$$\sum_{n \sim m} |a_n|^2 \le c \int_J |f(t)|^2 \, dt. \tag{9.28}$$

The lemma now follows from (9.27) and (9.28).

Next we establish a variant of Lemma 8.6 (p. 162):

Lemma 9.12. *Fix an interval I of length $|I| > 2\pi/\gamma$. There exists a sequence (φ_m) in $L^2(I)$ such that*

$$(\varphi_m, e_n)_{L^2(I)} = \delta_{mn} \quad \text{for all} \quad m, n. \tag{9.29}$$

Furthermore, the sequence (φ_m) is bounded in $L^2(I)$ by some constant depending only on M, γ, γ', and on the interval I.

Proof. Choose M intervals $I_1, \ldots, I_M$ of length $> 2\pi/(M\gamma)$ such that $I_1 + \cdots + I_M = I$. Combining the preceding lemma with Lemma 8.5 (p. 162), we see that for each $1 \le k \le M$ there exists in $L^2(I_k)$ a bounded sequence $(\varphi_{k,m})$ satisfying the conditions

$$(\varphi_{k,p}, e_q)_{L^2(I_k)} = \delta_{pq} \quad \text{whenever} \quad p \sim q, \tag{9.30}$$

and

$$(\varphi_{k,p}, f_q)_{L^2(I_k)} = 0 \quad \text{whenever} \quad p \not\sim q \text{ and } q \equiv k. \tag{9.31}$$

(The bounds depend only on γ, γ', and on the interval I.)

For every $m \in A_j$ set

$$\varphi_n = \varphi_{1,n} * \varphi_{2,m} * \varphi_{3,m} * \cdots * \varphi_{M,m}$$

for all $n \sim m$; then (φ_n) is a bounded sequence in $L^2(I)$ by elementary properties of the convolution. Furthermore, we have

$$\widehat{\varphi}_n(\lambda_q) = \widehat{\varphi}_{1,n}(\lambda_q) \widehat{\varphi}_{2,m}(\lambda_q) \cdots \widehat{\varphi}_{M,m}(\lambda_q) \tag{9.32}$$

for all integers q. If $q \not\sim m$, then choosing $1 \le k \le M$ such that $q \equiv k$, we see that the factor $\widehat{\varphi}_{1,n}(\lambda_q)$ (if $k = 1$) or $\widehat{\varphi}_{k,m}(\lambda_q)$ (if $k > 1$) on the right-hand side of (9.32) vanishes by (9.31), so that

$$(\varphi_n, f_q)_{L^2(I)} = \widehat{\varphi}_n(\lambda_q) = 0.$$

Taking linear combinations, it follows that

$$(\varphi_n, e_q)_{L^2(I)} = 0$$

whenever $q \not\sim m$.

It remains to prove that

$$(\varphi_n - \varphi_{1,n}, e_q)_{L^2(I)} = 0 \tag{9.33}$$

whenever $n \sim m$ and $q \sim n$. Indeed, then (9.29) will follow from (9.30).

Of course, (9.33) is equivalent to

$$(\varphi_n - \varphi_{1,n}, f_q)_{L^2(I)} = 0$$

for all $n \sim m$ and $q \sim n$, i.e., to

$$\widehat{\varphi}_n(\lambda_q) = \widehat{\varphi}_{1,n}(\lambda_q).$$

As a consequence of (9.32) this will follow if we show that

$$\widehat{\varphi}_{k,m}(\lambda_q) = 1$$

for all $2 \le k \le M$ and $q \sim m$. In fact, this is an easy consequence (even if $k = 1$) of the Newton interpolational formula

$$f_q = e_m + (\lambda_q - \lambda_m)e_{m+1} + \cdots + (\lambda_q - \lambda_m)\cdots(\lambda_q - \lambda_{q-1})e_q.$$

Indeed, we have

$$
\begin{aligned}
\widehat{\varphi}_{k,m}(\lambda_q) &= (\varphi_{k,m}, f_q) \\
&= (\varphi_{k,m}, e_m) + (\lambda_q - \lambda_m)(\varphi_{k,m}, e_{m+1}) + \cdots \\
&\qquad + (\lambda_q - \lambda_m)\cdots(\lambda_q - \lambda_{q-1})(\varphi_{k,m}, e_q) \\
&= (\varphi_{k,m}, e_m) = 1
\end{aligned}
$$

by the biorthogonality properties of $\varphi_{k,m}$.

9.5 Proof of the Inverse Part of Theorem 9.4

Let us observe that for every bounded interval I and for every number $\sigma > M - (1/2)$ we have clearly

$$\int_I |f(t)|^2 \, dt \le c \int_{-\infty}^{\infty} (1+t^2)^{-\sigma}|f(t)|^2 \, dt,$$

where c denotes the supremum of $(1+t^2)^\sigma$ on I. Hence it suffices to prove for every interval J of length $|J| > 2\pi/\gamma$ the estimate

$$\sum |a_n|^2 \le c \int_J \left| \sum a_n e_n \right|^2 \, dt. \tag{9.34}$$

We follow a similar strategy as in Section 8.4 (p. 161). Instead of establishing (9.34) directly for all finite sums

$$f = \sum a_n e_n, \tag{9.35}$$

it will be easier to construct a function y such that

$$\sum |a_n|^2 \le c(y, f)_{L^2(J)} \tag{9.36}$$

and

$$\|y\|_{L^2(J)}^2 \le c \sum |a_n|^2. \tag{9.37}$$

Then (9.34) will follow by applying the Cauchy–Schwarz inequality. Indeed, we have

$$\sum |a_n|^2 \le c(y, f)_{L^2(J)} \le c\|y\|_{L^2(J)} \cdot \|f\|_{L^2(J)} \le c\left(\sum |a_n|^2\right)^{1/2} \|f\|_{L^2(J)},$$

which implies (9.34).

Turning to the construction of y, choose an interval $I = (\alpha, \beta)$ of length $|I| > 2\pi/\gamma$ and a number $r > 0$ such that

$$J = (\alpha - r, \beta + r).$$

Then fix a biorthogonal sequence (φ_m) in $L^2(I)$ satisfying the conditions of the preceding lemma. Next, choose a real-valued even function $H \in C_c^\infty(-r/2, r/2)$ satisfying

$$\int_{-\infty}^\infty H(t)\, dt = 1$$

and

$$\int_{-\infty}^\infty t^j H(t)\, dt = 0, \quad j = 1, \dots, M,$$

and set $G := H * H$. Denoting by h and g the Fourier transforms of H and G, respectively, we have

$$g(\lambda) = |g(\lambda)|^2 \geq 0 \quad \text{for all } \lambda \in \mathbb{R}, \tag{9.38}$$
$$g(0) = 1, \tag{9.39}$$
$$g^{(j)}(0) = 0, \quad j = 1, \dots, M. \tag{9.40}$$

Fix a number $0 < \gamma' \leq \gamma$ such that

$$|g^{(j)}(\lambda)| \leq 1/M \quad \text{for all } |\lambda| \leq M\gamma' \text{ and } j = 1, \dots, M. \tag{9.41}$$

Now, given given a function f of the form (9.35), we define another function $y(t)$ by the formula

$$\widehat{g}(\lambda) = \sum a_n \widehat{\varphi}_n(\lambda) g(\lambda - \lambda_n).$$

We shall establish the estimates (9.36) and (9.37) in the following two lemmas. This will complete the proof of Theorem 9.4.

Lemma 9.13. *The estimates (9.36) are satisfied with a constant c that does not depend on the particular choice of f.*

Proof. Let us assume for the moment that if $m \in A_j$ for some $1 \leq j \leq M$, then

$$(y, e_p) = \sum_{q=m}^p g(\lambda_q - \lambda_q, \dots, \lambda_p - \lambda_q) a_q, \quad m \leq p < m + j, \tag{9.42}$$

where we use the divided differences of the function $g(\lambda)$.

As a consequence of (9.41) we deduce from the formula (9.5) that

$$|g(\lambda_q - \lambda_q, \dots, \lambda_p - \lambda_q)| \leq \frac{1}{M}$$

for all $m \leq q < p < m + j$. Since we have also $g(0) = 1$ by (9.39), we deduce from the formula

$$(y, f) = \sum_{j=1}^{M} \sum_{m \in A_j} \sum_{p \sim m} \sum_{q=m}^{p} g(\lambda_q - \lambda_q, \ldots, \lambda_p - \lambda_q) a_q \overline{a_p}$$

that

$$\sum_{j=1}^{M} \sum_{m \in A_j} \sum_{p \sim m} |a_p|^2 \leq |(y, f)| + \frac{1}{M} \sum_{j=1}^{M} \sum_{m \in A_j} \sum_{p \sim m} \sum_{q=m}^{p-1} |a_p| \cdot |a_q|$$

$$\leq |(y, f)| + \frac{1}{2} \sum_{j=1}^{M} \sum_{m \in A_j} \sum_{p \sim m} |a_p|^2.$$

Hence (9.36) follows with $c = 2$.

It remains to prove (9.42). By a continuity argument we may assume that the exponents $\lambda_m, \ldots, \lambda_{m+j-1}$ are distinct. Applying Newton's interpolational formula

$$f_r = \sum_{p=m}^{r} (\lambda_r - \lambda_m) \cdots (\lambda_r - \lambda_{p-1}) e_p, \quad m \leq r < m + j,$$

we deduce from (9.42) that

$$(y, f_r) = \sum_{p=m}^{r} (\lambda_r - \lambda_m) \cdots (\lambda_r - \lambda_{p-1}) \sum_{q=m}^{p} g(\lambda_q - \lambda_q, \ldots, \lambda_p - \lambda_q) a_q$$

$$= \sum_{q=m}^{r} \sum_{p=q}^{r} (\lambda_r - \lambda_m) \cdots (\lambda_r - \lambda_{p-1}) g(\lambda_q - \lambda_q, \ldots, \lambda_p - \lambda_q) a_q$$

$$= \sum_{q=m}^{r} (\lambda_r - \lambda_m) \cdots (\lambda_r - \lambda_{q-1}) \sum_{p=q}^{r} (\lambda_r - \lambda_q)$$

$$\cdots (\lambda_r - \lambda_{p-1}) g(\lambda_q - \lambda_q, \ldots, \lambda_p - \lambda_q) a_q$$

$$= \sum_{q=m}^{r} (\lambda_r - \lambda_m) \cdots (\lambda_r - \lambda_{q-1}) g(\lambda_r - \lambda_q) a_q,$$

i.e.,

$$(y, f_r) = \sum_{q=m}^{r} (\lambda_r - \lambda_m) \cdots (\lambda_r - \lambda_{q-1}) g(\lambda_r - \lambda_q) a_q, \quad m \leq r < m + j. \quad (9.43)$$

Now, since

$$(e_p)_{m \leq p < m+j} \quad \text{and} \quad (f_r)_{m \leq r < m+j}$$

form two bases of the same vector space by Lemma 9.5, (9.43) also implies (9.42). Hence it suffices to verify (9.43).

First of all, it follows from the orthogonality properties of φ_n that

$$\widehat{\varphi}_n(\lambda_r) = (\varphi_n, f_r) = 0$$

whenever $n \not\sim m$ and also if $r < n < m + j$. Therefore, we deduce from the definition of y that

$$(y, f_r) = \widehat{y}(\lambda_r) = \sum_{n=m}^{r} a_n \widehat{\varphi}_n(\lambda_r) g(\lambda_r - \lambda_n).$$

Comparing this with (9.43), we need to verify only the equalities

$$\widehat{\varphi}_n(\lambda_r) = (\lambda_r - \lambda_m) \cdots (\lambda_r - \lambda_{n-1}), \quad m \le n \le r < m + j,$$

and this follows from Newton's interpolational formula

$$\widehat{\varphi}_n(\lambda_r) = (\varphi_n, f_r) = \sum_{p=m}^{r} (\lambda_r - \lambda_m) \cdots (\lambda_r - \lambda_{p-1})(\varphi_n, e_p)$$

because $(\varphi_n, e_p) = \delta_{np}$.

Lemma 9.14. *The estimates (9.37) are satisfied with a constant c that does not depend on the particular choice of f.*

Proof. Applying Plancherel's theorem and using the nonnegativity of $g(\lambda)$, we have

$$\int_{-\infty}^{\infty} |y(t)|^2 \, dt = c \int_{-\infty}^{\infty} |\widehat{y}(\lambda)|^2 \, d\lambda$$

$$\le c \big(\sup_{n,\lambda} |\widehat{\varphi}_n(\lambda)|\big) \int_{-\infty}^{\infty} \Big(\sum_{n=-\infty}^{\infty} |a_n| g(\lambda - \lambda_n) \Big)^2 \, d\lambda.$$

Since the sequence (φ_n) is bounded in $L^2(I)$, we have uniformly in n and λ the estimate

$$|\widehat{\varphi}_n(\lambda)| \le \|\widehat{\varphi}_n\|_{L^\infty(\mathbb{R})} \le \|\varphi_n\|_{L^1(\mathbb{R})} = \|\varphi_n\|_{L^1(I)} \le \sqrt{|I|} \cdot \|\varphi_n\|_{L^2(I)} \le c.$$

Therefore, the above inequality implies that

$$\int_{-\infty}^{\infty} |y(t)|^2 \, dt \le c \int_{-\infty}^{\infty} \Big(\sum_{n=-\infty}^{\infty} |a_n| g(\lambda - \lambda_n) \Big)^2 \, dt. \tag{9.44}$$

We majorize the last integral by applying Plancherel's theorem again. Since

$$\sum_{n=-\infty}^{\infty} |a_n| g(\lambda - \lambda_n)$$

is the Fourier transform of

$$\sum_{n=-\infty}^{\infty} |a_n| e^{i\lambda_n t} G(t)$$

and since $G(t)$ vanishes outside the interval $(-r, r)$, we have

$$\int_{-\infty}^{\infty} \Big(\sum_{n=-\infty}^{\infty} |a_n| g(\lambda - \lambda_n) \Big)^2 \, d\lambda = c \int_{-r}^{r} \Big| \sum_{n=-\infty}^{\infty} |a_n| e^{i\lambda_n t} G(t) \Big|^2 \, dt.$$

Since G is continuous and has a compact support, it is bounded. Therefore, the last expression is majorized by

$$c \int_{-r}^{r} \Big| \sum_{n=-\infty}^{\infty} |a_n| e^{i\lambda_n t} \Big|^2 \, dt,$$

so that we deduce from (9.44) the inequality

$$\int_{-\infty}^{\infty} |y(t)|^2 \, dt \le c \int_{-r}^{r} \Big| \sum_{n=-\infty}^{\infty} |a_n| e^{i\lambda_n t} \Big|^2 \, dt.$$

It remains to show that

$$\int_{-r}^{r} \Big| \sum_{n=-\infty}^{\infty} |a_n| e^{i\lambda_n t} \Big|^2 \, dt \le c \sum_{n=-\infty}^{\infty} |a_n|^2.$$

This follows by decomposing the sequence into M subsequences and applying Ingham's theorem to each corresponding integral. (This is justified by condition (9.2).) We have

$$\int_{-r}^{r} \Big| \sum_{n=-\infty}^{\infty} |a_n| e^{i\lambda_n t} \Big|^2 \, dt \le M \sum_{k=1}^{M} \int_{-r}^{r} \Big| \sum_{n \equiv k} |a_n| e^{i\lambda_n t} \Big|^2 \, dt$$

$$\le M \sum_{k=1}^{M} c_k \sum_{n \equiv k} |a_n|^2$$

$$\le c \sum_{n=-\infty}^{\infty} |a_n|^2.$$

9.6 Singular Points of Dirichlet Series

Consider a Dirichlet series of the type

$$f(s) = f(\sigma + ti) = \sum_{n=1}^{\infty} d_n e^{-\lambda_n s}, \tag{9.45}$$

in which (λ_n) is a strictly increasing sequence of strictly positive real numbers, σ and t denote the real and imaginary parts of s, respectively, and the coefficients d_n are given real or complex numbers. We recall from the general theory (see, e.g., [13]) that there exists a number $-\infty \leq \sigma_0 \leq \infty$ such that the series converges whenever $\sigma > \sigma_0$, and diverges whenever $\sigma < \sigma_0$. If this *convergence abscissa* σ_0 is finite, then the vertical line $\sigma = \sigma_0$ is called the *convergence line* of the series (9.45).

We are going to give a new proof of the following theorem of Bernstein [13]:

Theorem 9.15. *Assume that the series* (9.45) *has a finite convergence abscissa, and that the sequence* (λ_n) *has a finite upper density* D^+. *Furthermore, assume that*

$$\frac{\log \min\{\lambda_{n+1} - \lambda_n, 1\}}{\lambda_n} \to 0 \qquad (9.46)$$

as $n \to \infty$. *Then every segment of length* $2\pi D^+$ *on the convergence line contains at least one singular point of* $f(s)$.

Remark. Bernstein's theorem generalized an earlier one due to Pólya [113]. Pólya assumed instead of the finiteness of D^+ and (9.46) the stronger uniform gap condition

$$\lambda_{n+1} - \lambda_n \geq \gamma \quad \text{for all} \quad n$$

for some $\gamma > 0$, and he concluded that every segment of length $2\pi/\gamma$ on the convergence line contains at least one singular point of $f(s)$. Ingham gave a simpler proof by applying his Theorem 4.3 (p. 59). We are going to proceed in a similar way to prove Bernstein's theorem by applying Theorem 9.4 (p. 177).

Assuming that Theorem 9.15 is false, we may assume without loss of generality that $D^+ = 1$, that the convergence abscissa is equal to 0, and that $f(s)$ has no singularity on the segment $\sigma = 0$ and $-\pi \leq t \leq \pi$.

We may then choose a sufficiently small $\varepsilon > 0$ such that $f(s)$ is regular in the closed rectangle

$$-3\varepsilon \leq \sigma \leq 1 + 3\varepsilon, \quad |t| \leq \pi + 4\varepsilon;$$

in particular, it is bounded by some constant A in this rectangle. For every nonnegative integer q and for every $\sigma > 0$ we have the convergent expansion

$$f^{(q)}(s) = \sum_{m=1}^{\infty} (-1)^q d_m \lambda_m^q e^{-\lambda_m s}.$$

For

$$0 < \sigma \leq 1, \quad |t| \leq \pi + \varepsilon,$$

applying Cauchy's formulae we have, denoting by C the circle $|z| = 3\varepsilon$, the following estimates:

$$|f^{(q)}(s)| = \left| \frac{q!}{2\pi i} \int_C \frac{f(s+z)}{z^{q+1}} \, dz \right| \le \frac{q!A}{(3\varepsilon)^q}.$$

Putting

$$b_m = (-1)^q d_m \lambda_m^q e^{-\lambda_m \sigma},$$

we may write

$$f^{(q)}(s) = \sum_{m=1}^{\infty} b_m e^{-i\lambda_m t}.$$

We are going to estimate the coefficients b_m. Let us choose γ and M by applying Proposition 9.3 (p. 175) to the interval $I := (-\pi - \varepsilon, \pi + \varepsilon)$. Choosing furthermore $\gamma' := \min\{\gamma, 1\}$ and introducing the divided differences as in Section 9.1 (p. 177), we may rewrite the above function in the form

$$f^{(q)}(s) = \sum_{n=1}^{\infty} a_n e_n(t).$$

Applying Theorem 9.4 (p. 177) to this series, we obtain the inequalities

$$|a_n| \le C(\varepsilon) \frac{q!A}{(3\varepsilon)^q}$$

with some constant $C(\varepsilon)$ independent of n and q.

Putting

$$P_m := \prod_{j=n_k}^{n_{k+1}-2} (\lambda_{j+1} - \lambda_j) \quad \text{for} \quad n_k \le m < n_{k+1}$$

(with $P_m := 1$ if $n_{k+1} = n_k + 1$), we have

$$|P_{m,n}| \ge P_m (\gamma')^{n-n_{k+1}} \ge P_m (\gamma')^{-M}$$

for all $n_k \le m \le n < n_{k+1}$. Therefore, using the relations (3.4) between the coefficients a_n and b_m, we deduce from the above estimates of a_n that

$$|b_m| \le \frac{M(\gamma')^M C(\varepsilon) A}{P_m} \frac{q!}{(3\varepsilon)^q} =: \frac{B}{P_m} \frac{q!}{(3\varepsilon)^q}.$$

More explicitly, we have

$$|(-1)^q d_m \lambda_m^q e^{-\lambda_m \sigma}| \le \frac{B}{P_m} \frac{q!}{(3\varepsilon)^q}.$$

Letting $\sigma \to 0$, it follows that

$$|d_m| \frac{(2\lambda_m \varepsilon)^q}{q!} \le \frac{B}{P_m} \left(\frac{2}{3}\right)^q.$$

Summing from $q = 0$ to ∞, we deduce that

$$|d_m|e^{2\lambda_m \varepsilon} \leq \frac{3B}{P_m}.$$

Hence

$$|d_m|e^{\lambda_m \varepsilon} \leq \frac{3B}{P_m}e^{-\lambda_m \varepsilon}.$$

We will show that the right-hand side of this inequality is of order $O(e^{-m\gamma\varepsilon/2})$ as $m \to \infty$. This will imply that the Dirichlet series converges for $s = -\varepsilon$, which is impossible because $-\varepsilon$ is strictly smaller than the convergence abscissa $\sigma = 0$.

It remains to establish the estimate

$$\frac{e^{-\lambda_m \varepsilon}}{P_m} \leq C'e^{-m\gamma\varepsilon/2}$$

for all sufficiently large m, with some constant C'. Here and in the sequel, C' denotes various constants independent of m.

Taking logarithms, we see that the desired estimate is equivalent to

$$-\lambda_m \varepsilon + \log \frac{1}{P_m} \leq C' - \frac{m\gamma\varepsilon}{2},$$

or to

$$\frac{m\gamma\varepsilon}{2\lambda_m} \leq \varepsilon - \frac{1}{\lambda_m}\log\frac{1}{P_m} + \frac{C'}{\lambda_m}. \tag{9.47}$$

Here the left-hand side is majorized by an expression converging to $\varepsilon/2$. Indeed, observe that writing $m = Mk + r$ with integers $k \geq 0$ and $1 \leq r \leq M$, we have

$$\lambda_m \geq Mk\gamma + \lambda_r = (m-r)\gamma + \lambda_r \geq (m-M)\gamma.$$

Therefore,

$$\frac{m\gamma\varepsilon}{2\lambda_m} \leq \frac{m\gamma\varepsilon}{2(m-M)\gamma} = \frac{\varepsilon}{2}\frac{m}{m-M} \to \frac{\varepsilon}{2}$$

as $m \to \infty$.

We complete our proof by showing that the right-hand side of (9.47) converges to ε. Since $\lambda_m \to \infty$, we have only to show that

$$\frac{\log P_m}{\lambda_m} \to 0.$$

Since

$$\left|\frac{\log P_m}{\lambda_m}\right| = \sum_{j=n_k}^{n_{k+1}-2}\left|\frac{\log(\lambda_{j+1} - \lambda_j)}{\lambda_m}\right| = \sum_{j=n_k}^{n_{k+1}-2}\left|\frac{\log(\lambda_{j+1} - \lambda_j)}{\lambda_j}\frac{\lambda_j}{\lambda_m}\right|$$

if $n_k \leq m < n_{k+1}$, the desired relation follows from our hypothesis (9.46) because the number of terms in this sum is bounded by M and

$$0 < \frac{\lambda_j}{\lambda_m} < \frac{\lambda_{n_{k+1}}}{\lambda_{n_k}} \to 1.$$

Remark. According to terminology introduced by Bernstein [13], the relation

$$\frac{1}{\lambda_m} \log \frac{1}{P_m} \to 0$$

means that the *condensation index* of the sequence (λ_m) is equal to zero.

10

Problems with Weakened Gap Conditions

In this chapter we demonstrate the usefulness of the general theorem of the preceding chapter by proving optimal simultaneous observability results for string and beam systems. Here we also need some tools from Diophantine approximation: generally speaking, our results hold only under some number-theoretic hypotheses concerning the lengths of the strings or beams. The set of exceptional parameters has zero Lebesgue measure, but we prove that it has a maximal Hausdorff dimension. At the end we establish an optimal observability theorem for spherical shells with a central hole.

10.1 Simultaneous Observability of a System of Strings

Consider a finite number of vibrating strings with fixed endpoints, one of which is common to all of them. Assuming that we can observe only the *combined* force exerted by the strings at the common endpoint during some time T, it is natural to ask whether we can identify all initial data.

With ℓ_j denoting the lengths of the strings, a reasonable model describing these vibrations is given by the following *uncoupled system*:

$$\begin{cases} u_{j,tt} - u_{j,xx} + a_j u_j = 0 & \text{in } \mathbb{R} \times (0, \ell_j), \\ u_j(t,0) = u_j(t, \ell_j) = 0 & \text{for } t \in \mathbb{R}, \\ u_j(0,x) = u_{j0}(x) \quad \text{and} \quad u_{j,t}(0,x) = u_{j1}(x) & \text{for } x \in (0, \ell_j), \\ j = 1, \ldots, N, \end{cases} \tag{10.1}$$

where $N \geq 2$ and $a_1, \ldots, a_N$ are given real numbers. We would like to know whether the linear map

$$(u_{10}, \ldots, u_{N0}, u_{11}, \ldots, u_{N1}) \mapsto \sum_{j=1}^{N} u_{j,x}(\cdot, 0) \tag{10.2}$$

is one-to-one in suitable, natural functions spaces. If this is so, then we would also like to obtain more precise, quantitative norm estimates.

This problem was first studied by Jaffard, Tucsnak, and Zuazua [57] for $N = 2$.

It follows from Propositions 4.7 and 4.8 (p. 73) that the system (10.1) is well-posed for

$$u_{j0} \in H_0^1(0, \ell_j) \quad \text{and} \quad u_{j1} \in L^2(0, \ell_j), \quad j = 1, \dots, N,$$

and that formula (10.2) defines a continuous linear map of

$$\left(\prod_{j=1}^N H_0^1(0, \ell_j) \right) \times \left(\prod_{j=1}^N L^2(0, \ell_j) \right)$$

into $L_{loc}^2(\mathbb{R})$. More precisely, the solutions of (10.1) satisfy the estimates

$$\int_I \left| \sum_{j=1}^N u_{j,x}(t, 0) \right|^2 dt \le c_I \sum_{j=1}^N \left(\int_\Omega |\nabla u_{j0}|^2 + |u_{j1}|^2 \, dx \right)$$

on every interval I.

Introducing the Hilbert spaces D^s as in Section 3.2 (p. 38), this estimate can be rewritten in the form

$$\int_I \left| \sum_{j=1}^N u_{j,x}(t, 0) \right|^2 dt \le c_I \sum_{j=1}^N \left(\|u_{j0}\|_1^2 + \|u_{j1}\|_0^2 \right).$$

We are going to establish the following weakened converse of this inequality:

Proposition 10.1. *For almost all choices of $(\ell_1, \dots, \ell_N) \in (0, \infty)^N$, the solutions of (10.1) satisfy the estimates*

$$\sum_{j=1}^N \left(\|u_{j0}\|_s^2 + \|u_{j1}\|_{s-1}^2 \right) \le c_{s,I} \int_I \left| \sum_{j=1}^N u_{j,x}(t, 0) \right|^2 dt \tag{10.3}$$

on every interval I of length

$$|I| > 2(\ell_1 + \cdots + \ell_N), \tag{10.4}$$

for every $s < 2 - N$.

Moreover, if the numbers a_j are distinct, then the estimate (10.3) also holds in the limiting case $s = 2 - N$.

Remarks.

- The optimal condition (10.4) was first given (even for the case $N = 2$) in [8], with the proof published in [9].

- If no lower-order terms are present in the equation, i.e., if $a_1 = \cdots = a_N = 0$, then the proposition can be proved by applying D'Alembert's formula; see Dáger and Zuazua [28]. We will show in this section that our first proof, given in [8], [9], easily extends to the general case.
- The necessity of a condition on the lengths of the strings is shown already in the case $N = 2$ by the following example, given in [7]: if $\ell_1/\ell_2 = p/q$ is a rational number, then the map (10.2) is not one-to-one for any interval I. Indeed, the formulae

$$u_1(t, x) = e^{ip\pi t/\ell_1} \sin(p\pi x/\ell_1),$$
$$u_2(t, x) = -e^{iq\pi t/\ell_2} \sin(q\pi x/\ell_2),$$

define a nonzero solution of (10.1) for which the right-hand side of (10.2) vanishes on $\mathbb{R}$.

For our proof we recall, e.g., from [22], the following classical result of Diophantine approximation:

Proposition 10.2. *There exists a set $Q \subset \mathbb{R}$ of zero Lebesgue measure such that if $z \in \mathbb{R} \backslash Q$, then*

$$\mathrm{dist}\,(kz, \mathbb{Z}) \geq \frac{c_\alpha(z)}{k^\alpha}, \quad k = 1, 2, \ldots,$$

for every $\alpha > 1$.

It is clear that Q contains the rational numbers. In the sequel we assume that the lengths of the strings satisfy the condition

$$\ell_j/\ell_m \notin Q \quad \text{for all} \quad j \neq m. \tag{10.5}$$

Using Fubini's theorem one can readily verify that almost all N-tuples $(\ell_1, \ldots, \ell_N) \in (0, \infty)^N$ satisfy this condition.

Let us also recall from the proof of Propositions 3.2 and 4.7 that putting

$$\mu_{j,k} := k\pi/\ell_j \quad \text{and} \quad \omega_{j,k} := \sqrt{\mu_{j,k}^2 + a_j}$$

for brevity, the solutions of (10.1) are given by the formulae

$$u_j(t, x) = \sum_{k=1}^{\infty} (b_{j,k} e^{i\omega_{j,k} t} + b_{j,-k} e^{-i\omega_{j,k} t}) \sin \mu_{j,k} x \tag{10.6}$$

with suitable complex coefficients $b_{j,k}$ and $b_{j,-k}$. More precisely, these formulae are valid except for countably many particular values of the lengths ℓ_j, where $\omega_{j,k} = 0$ for some k. Since the corresponding N-tuples $(\ell_1, \ldots, \ell_N)$ form a set of zero Lebesgue measure, we may exclude them from the considerations that follow.

We need the following auxiliary result for the proof of Proposition 10.1.

Lemma 10.3. *Assume* (10.5). *There exists a number* $\gamma > 0$ *such that if*

$$0 < |\omega_{j,k} - \omega_{m,n}| \le \gamma, \tag{10.7}$$

then $j \neq m$ *and*

$$|\omega_{j,k} - \omega_{m,n}| \ge \frac{c_\alpha}{|\omega_{j,k}|^\alpha} \quad \text{and} \quad |\omega_{j,k} - \omega_{m,n}| \ge \frac{c_\alpha}{|\omega_{m,n}|^\alpha}$$

for every $\alpha > 1$, *with a constant* c_α *independent of the particular choice of* $\omega_{j,k}$ *and* $\omega_{m,n}$.

Moreover, if the numbers a_j are distinct, then we even have the estimates

$$|\omega_{j,k} - \omega_{m,n}| \ge \frac{c_\alpha}{|\omega_{j,k}|} \quad \text{and} \quad |\omega_{j,k} - \omega_{m,n}| \ge \frac{c_\alpha}{|\omega_{m,n}|}.$$

Proof. Let us first consider the case $a_1 = \cdots = a_N = 0$. Then we have

$$\omega_{j,k} = \mu_{j,k} := k\pi/\ell_j,$$

so that if we choose $\gamma < \min_j \pi/\ell_j$, the condition (10.7) implies that $j \neq m$. Note that this condition also implies the asymptotic relations

$$|\omega_{j,k}| \asymp |\omega_{m,n}| \asymp k \asymp n. \tag{10.8}$$

Furthermore, applying Proposition 10.2, it follows from (10.7) that

$$|\omega_{j,k} - \omega_{m,n}| \asymp |k(\ell_m/\ell_j) - n| \ge \frac{c_\alpha}{k^\alpha},$$

and the lemma follows by using (10.8).

Turning to the general case, first we note the obvious asymptotic relations

$$\omega_{j,k} = \frac{k\pi}{\ell_j} + \frac{a_j \ell_j}{2k\pi} + O\left(\frac{1}{k^3}\right), \quad k \to \infty,$$

for every $j = 1, \ldots, N$. Hence, by choosing a possibly smaller value of γ, we have that (10.7) still implies the relation $j \neq m$.

If (10.7) is satisfied, then these relations also imply that

$$\begin{aligned}
\frac{a_j \ell_j}{2k\pi} - \frac{a_m \ell_m}{2n\pi} &= \frac{\ell_j}{2k\pi}\left(a_j - a_m \frac{k/\ell_j}{n/\ell_m}\right) \\
&= \frac{\ell_j}{2k\pi}\left(a_j - a_m \frac{k/\ell_j}{(k/\ell_j) + O(1)}\right) \\
&= \frac{\ell_j}{2k\pi}\left(a_j - a_m(1 - O(1/k))\right).
\end{aligned}$$

Hence

$$\left|\frac{a_j \ell_j}{2k\pi} - \frac{a_m \ell_m}{2n\pi}\right| = \frac{|a_j - a_m|\ell_j}{2k\pi} + O\left(\frac{1}{k^2}\right).$$

If $a_j = a_m$, then it follows that

$$|\omega_{j,k} - \omega_{m,n}| \geq \frac{c_\alpha}{k^\alpha} - \frac{c'}{k^2} \geq \frac{c'_\alpha}{k^\alpha}$$

for every $1 < \alpha < 2$. Of course, for greater values of α the final estimate still holds, because then k^α grows even faster.

If $a_j \neq a_m$, then we have simply

$$|\omega_{j,k} - \omega_{m,n}| \geq \frac{|a_j - a_m|\ell_j}{2k\pi} - O\left(\frac{1}{k^2}\right) - O\left(\frac{1}{k^\alpha}\right) \geq \frac{c}{k}.$$

Proof of Proposition 10.1. Using the representation (10.6) of the solutions and using the relations $\omega_{j,k} \asymp k$, the estimates (10.3) can be rewritten in the following equivalent form:

$$\sum_{j=1}^{N} \sum_{k=1}^{\infty} k^{2s} \left(|b_{j,k}|^2 + |b_{j,-k}|^2\right)$$

$$\leq c_{s,I} \int_I \left| \sum_{j=1}^{N} \sum_{k=1}^{\infty} \mu_{j,k} b_{j,k} e^{i\omega_{j,k}t} + \mu_{j,k} b_{j,-k} e^{-i\omega_{j,k}t} \right|^2 dt.$$

Let us rearrange the exponents $\pm\omega_{j,k}$ into an increasing sequence λ_n. Excluding a set of zero measure of the N-tuples $(\ell_1, \ldots, \ell_N)$, we may assume that $\lambda_n \neq \lambda_k$ whenever $n \neq k$. Denoting the corresponding coefficients $\mu_{j,k} b_{j,\pm k}$ by b_n for brevity and using the relations $\mu_{j,k} \asymp k$, we have to establish the following estimate:

$$\sum_{n=-\infty}^{\infty} |\lambda_n|^{2s-2} |b_n|^2 \leq c_{s,I} \int_I \left| \sum_{n=-\infty}^{\infty} b_n e^{i\lambda_n t} \right|^2 dt. \tag{10.9}$$

It follows from the structure of the exponents λ_n that

$$\sum_{j=1}^{N} \left(\frac{r\ell_j}{\pi} - 1\right) \leq n^+(r) \leq \sum_{j=1}^{N} \left(\frac{r\ell_j}{\pi} + 1\right)$$

for all $r > 0$, whence

$$D^+ = (\ell_1 + \cdots + \ell_N)/\pi.$$

Now, given a bounded interval I of length $> 2(\ell_1 + \cdots + \ell_N) = 2\pi D^+$, choose γ, M as in Proposition 9.3 (p. 175) and then choose γ' such that $N\gamma' < \pi/\ell_j$ for all j. Then no chain of close exponents is longer than N; i.e., $A_j = \emptyset$ for all $j > N$. Introducing the functions $e_n(t)$ as in Theorem 9.4 (p. 177), we have the estimates

$$\int_I \left| \sum_{n=-\infty}^{\infty} a_n e_n(t) \right|^2 dt \asymp \sum_{n=-\infty}^{\infty} |a_n|^2. \tag{10.10}$$

If $m \in A_j$ for some $j = 1, \ldots, N$, then rewriting the sums according to the equation

$$\sum_{n=m}^{m+j-1} b_n e^{i\lambda_n t} = \sum_{n=m}^{m+j-1} a_n e_n(t),$$

we obtain from the structure of the divided differences the inequality

$$\sum_{n=m}^{m+j-1} |b_n|^2 \cdot \min\{|\lambda_p - \lambda_q| \; : \; m \le p < q < m+j\}^{2N-2} \le C \sum_{n=m}^{m+j-1} |a_n|^2$$

with a constant C independent of the choice of m. Furthermore, applying Lemma 10.3, we have

$$\min\{|\lambda_p - \lambda_q| \; : \; m \le p < q < m+j\}^{2N-2} \ge C_\alpha \max\{|\lambda_n| \; : \; m \le n < m+j\}^\beta$$

for every $\beta < 2 - 2N$. Therefore, we deduce from the last inequality that

$$\sum_{n=m}^{m+j-1} |\lambda_n|^\beta |b_n|^2 \le C_\beta \sum_{n=m}^{m+j-1} |a_n|^2$$

for all $\beta < 2 - 2N$, and then

$$\sum_{n=-\infty}^{\infty} |\lambda_n|^\beta |b_n|^2 \le C_\beta \sum_{n=-\infty}^{\infty} |a_n|^2. \tag{10.11}$$

We combine (10.10) and (10.11) and observe that the condition $2s - 2 < 2 - 2N$ is equivalent to $s < 2 - N$. The estimate (10.9) follows.

10.2 The Hausdorff Dimension of the Set of Exceptional Parameters

In Proposition 10.1 of the preceding chapter we excluded an exceptional set of zero Lebesgue measure of the N-tuples $(\ell_1, \ldots, \ell_N)$. We show in this section that this set is in fact large in the sense that its *Hausdorff dimension* is maximal.

Let us first recall the definition of the Hausdorff dimension; we refer, e.g., to Falconer [33] for proofs. Given a set $F \subset \mathbb{R}^N$ and positive numbers s and ε, set

$$\mu_\varepsilon^s(F) := \inf_{\mathcal{B}} \sum_{B \in \mathcal{B}} (\operatorname{diam} B)^s,$$

where $\mathcal{B}$ runs over all countable covers of F by sets of diameter $\le \varepsilon$.

It follows from the definition of the infimum that $\mu_\varepsilon^s(F)$ can only increase as $\varepsilon \to 0$, so that we may also define

$$\mu^s(F) := \lim_{\varepsilon \to 0} \mu_\varepsilon^s(F) = \sup_{\varepsilon > 0} \mu_\varepsilon^s(F).$$

It is called the *s-dimensional Hausdorff outer measure* of F.

Next one can prove that there exists a critical value $0 \le s_0 \le \infty$, called the *Hausdorff dimension* of F, such that

$$\mu^s(F) = \infty \quad \text{for all} \quad s < s_0$$

and

$$\mu^s(F) = 0 \quad \text{for all} \quad s > s_0.$$

Examples.

- Every nonempty open subset of $\mathbb{R}^N$ is N-dimensional.
- The Hausdorff dimension of the triadic Cantor set is equal to $\ln 2 / \ln 3 \approx 0.63$.

For every fixed $\varepsilon > 0$, let us denote by Q_ε the set of real numbers a for which there exists a sequence $p_1 < p_2 < \cdots$ of positive integers sayisfying the asymptotic relations

$$\text{dist}\, (p_k a, \mathbb{Z}) = o\big(p_k^{-1-\varepsilon}\big), \quad k \to \infty.$$

It follows from Proposition 10.2 that Q_ε has zero Lebesgue measure for every $\varepsilon > 0$. But its Hausdorff dimension is positive: more precisely, we have the following easy consequence of a classical result due to Jarník [58] (see also Theorem 8.16 in [33]):

Proposition 10.4. *The Hausdorff dimension of Q_ε is equal to $2/(2+\varepsilon)$.*

Returning to our problem, for simplicity we restrict ourselves to the case $N = 2$ with $a_1 = a_2 = 0$. Using Proposition 10.4 we are going to establish the following result:

Proposition 10.5. *Consider the system* (10.1) *with $N = 2$ and $a_1 = a_2 = 0$, and fix an arbitrary interval I. For each $\varepsilon > 0$, the pairs $(\ell_1, \ell_2) \in (0, \infty)^2$ for which the estimate*

$$\sum_{j=1}^{2} \|u_{j0}\|_{-\varepsilon}^2 + \|u_{j1}\|_{-\varepsilon-1}^2 \le c_{\varepsilon,I} \int_I |u_{1,x}(0,t) + u_{2,x}(0,t)|^2 \, dt \qquad (10.12)$$

holds for all solutions of (10.1) *form a set of Hausdorff dimension at least equal to*

$$2 - \frac{\varepsilon}{2 + \varepsilon}.$$

Proof. It suffices to show that if $a := \ell_1/(\ell_1 + \ell_2) \in Q_\varepsilon$, then the estimate (10.12) does not hold for all solutions of (10.1). Indeed, since the map

$$x \mapsto \frac{x}{1-x},$$

which transforms $\ell_1/(\ell_1 + \ell_2)$ into ℓ_1/ℓ_2, does not diminish the distances, it follows that the set of the corresponding fractions ℓ_1/ℓ_2 has Hausdorff dimension at least equal to $2/(2+\varepsilon)$. Then an elementary argument shows that the corresponding pairs (ℓ_1, ℓ_2) form a set of Hausdorff dimension

$$\geq 2 - \frac{\varepsilon}{2+\varepsilon}.$$

If $a \in Q_\varepsilon$, then there exists a sequence of positive integers p_k, tending to infinity, such that

$$\text{dist}\,(p_k a, \mathbb{Z}) = o\big(p_k^{-1-\varepsilon}\big), \quad k \to \infty.$$

Choose positive integers n_k such that

$$|p_k a - n_k| = o\big(p_k^{-1-\varepsilon}\big);$$

then setting $m_k := p_k - n_k$, we have

$$|m_k a - n_k(1-a)| = o\big(p_k^{-1-\varepsilon}\big)$$

and

$$m_k \asymp n_k \asymp p_k$$

as $k \to \infty$, so that

$$\left|\frac{m_k}{\ell_2} - \frac{n_k}{\ell_1}\right| = o\big(n_k^{-1-\varepsilon}\big) = o\big(m_k^{-1-\varepsilon}\big).$$

Now for each fixed k, the formulae

$$u_1(t,x) = e^{in_k \pi t/\ell_1} \sin(n_k \pi x/\ell_1),$$
$$u_2(t,x) = -e^{im_k \pi t/\ell_2} \sin(m_k \pi x/\ell_2),$$

define a solution of (10.1), whose initial data satisfy the estimate

$$\sum_{j=1}^{2} \|u_{j0}\|_{-\varepsilon}^2 + \|u_{j1}\|_{-\varepsilon-1}^2 \asymp n_k^{-2\varepsilon}. \tag{10.13}$$

Furthermore, we have

$$
\begin{aligned}
u_{1,x}(t,0) + u_{2,x}(t,0) &= \frac{n_k \pi}{\ell_1} e^{in_k \pi t/\ell_1} - \frac{m_k \pi}{\ell_2} e^{im_k \pi t/\ell_2} \\
&= \left(\frac{n_k}{\ell_1} - \frac{m_k}{\ell_2}\right)\pi e^{in_k \pi t/\ell_1} + \frac{m_k \pi}{\ell_2}\left(e^{in_k \pi t/\ell_1} - e^{im_k \pi t/\ell_2}\right) \\
&= o\big(n_k^{-1-\varepsilon}\big) + m_k O\left(\frac{n_k}{\ell_1} - \frac{m_k}{\ell_2}\right) \\
&= o\big(n_k^{-\varepsilon}\big).
\end{aligned}
$$

Hence for every bounded interval I we have the estimate

$$\int_I |u_{1,x}(t,0) + u_{2,x}(t,0)|^2 \ dt = o(n_k^{-2\varepsilon}). \qquad (10.14)$$

Comparing (10.13) and (10.14), we conclude that (10.12) does not hold.

Remark. It follows from the proposition that the set of pairs (ℓ_1, ℓ_2) for which the estimate (10.12) fails for some $\varepsilon > 0$ has full Hausdorff dimension 2.

10.3 Simultaneous Observability of a System of Beams

In this section we investigate a similar problem as in Section 10.1, but for beams instead of strings. Given a positive integer $N \geq 2$ and positive numbers $\ell_1, \ldots, \ell_N$, we consider the solutions of the following *uncoupled system*:

$$\begin{cases} u_{j,tt} + u_{j,xxxx} = 0 & \text{in} \quad \mathbb{R} \times (0, \ell_j), \\ u_j(t,0) = u_j(t, \ell_j) = 0 & \text{for} \quad t \in \mathbb{R}, \\ u_{j,xx}(t,0) = u_{j,xx}(t, \ell_j) = 0 & \text{for} \quad t \in \mathbb{R}, \qquad (10.15) \\ u_j(0,x) = u_{j0}(x) \quad \text{and} \quad u_{j,t}(0,x) = u_{j1}(x) & \text{for} \quad x \in (0, \ell_j), \\ j = 1, \ldots, N. \end{cases}$$

We investigate again the nature of the linear map

$$(u_{10}, \ldots, u_{N0}, u_{11}, \ldots, u_{N1}) \mapsto \sum_{j=1}^{N} u_{j,x}(\cdot, 0). \qquad (10.16)$$

This problem was first studied in [7] in the special case $N = 2$, without using the main Theorem 9.4 of the preceding chapter. The following more general result and its proof are due to Sikolya [125].

It follows from Propositions 5.3 and 5.4 (p. 85) that the system (10.15) is well-posed for

$$u_{j0} \in H_0^1(0, \ell_j) \quad \text{and} \quad u_{j1} \in H^{-1}(0, \ell_j), \quad j = 1, \ldots, N,$$

and that formula (10.16) defines a continuous linear map of

$$\left(\prod_{j=1}^{N} H_0^1(0, \ell_j) \right) \times \left(\prod_{j=1}^{N} H^{-1}(0, \ell_j) \right)$$

into $L_{\text{loc}}^2(\mathbb{R})$. More precisely, introducing the Hilbert spaces D^s as in Section 3.2 (p. 42), we have

$$\int_I \left| \sum_{j=1}^N u_{j,x}(t,0) \right|^2 dt \le c_I \sum_{j=1}^N \left(\|u_{j0}\|_1^2 + \|u_{j1}\|_{-1}^2 \right).$$

We are going to establish the following weakened converse of this inequality:

Proposition 10.6. *For almost all choices of* $(\ell_1,\ldots,\ell_N) \in (0,\infty)^N$, *the solutions of* (10.15) *satisfy the estimates*

$$\sum_{j=1}^N \left(\|u_{j0}\|_s^2 + \|u_{j1}\|_{s-2}^2 \right) \le c_{s,I} \int_I \left| \sum_{j=1}^N u_{j,x}(t,0) \right|^2 dt \qquad (10.17)$$

on every interval I *and for every* $s < 1$.

Remark. The necessity of a condition on the lengths of the beams is shown by the following example for $N = 2$: if $\ell_1/\ell_2 = p/q$ is a rational number, then the map (10.16) is not one-to-one for any interval I. Indeed, the formulae

$$u_1(t,x) = e^{ip^2\pi^2 t/\ell_1^2} \sin(p\pi x/\ell_1),$$
$$u_2(t,x) = -e^{iq^2\pi^2 t/\ell_2^2} \sin(q\pi x/\ell_2),$$

define a nonzero solution of (10.15) for which the right-hand side of (10.16) vanishes on $\mathbb{R}$.

We recall from the proof of Propositions 5.3 and 5.4 that with

$$\mu_{j,k} := k\pi/\ell_j \quad \text{and} \quad \omega_{j,k} := \mu_{j,k}^2$$

for brevity, the solutions of (10.15) are given, apart from a set of N-tuples $(\ell_1,\ldots,\ell_N)$ of measure zero, by the formulae

$$u_j(t,x) = \sum_{k=1}^\infty (b_{j,k} e^{i\omega_{j,k}t} + b_{j,-k} e^{-i\omega_{j,k}t}) \sin \mu_{j,k} x \qquad (10.18)$$

with suitable complex coefficients $b_{j,k}$ and $b_{j,-k}$.

We need the following auxiliary result for the proof of Proposition 10.6, where we use the set Q introduced in Proposition 10.2 (p. 201).

Lemma 10.7. *Assume that*

$$\ell_j/\ell_m \notin Q \quad \text{for all} \quad j \ne m. \qquad (10.19)$$

Then there exists a number $\gamma > 0$ *such that if*

$$0 < |\omega_{j,k} - \omega_{m,n}| \le \gamma, \qquad (10.20)$$

then $j \ne m$ *and*

$$|\omega_{j,k} - \omega_{m,n}| \ge \frac{c_\beta}{|\omega_{j,k}|^\beta} \quad \text{and} \quad |\omega_{j,k} - \omega_{m,n}| \ge \frac{c_\beta}{|\omega_{m,n}|^\beta}$$

for every $\beta > 0$, *with a constant* c_β *independent of the particular choice of* $\omega_{j,k}$ *and* $\omega_{m,n}$.

Proof. We have
$$\omega_{j,k} = \mu_{j,k}^2 := k^2\pi^2/\ell_j^2,$$
so that with $\gamma < \min_j \pi^2/\ell_j^2$, the condition (10.20) implies that $j \neq m$. Note that this condition also implies the asymptotic relations
$$|\omega_{j,k}| \asymp |\omega_{m,n}| \asymp k^2 \asymp n^2. \tag{10.21}$$

Furthermore, applying Proposition 10.2, it follows from (10.20) and (10.21) that
$$\begin{aligned}
|\omega_{j,k} - \omega_{m,n}| &= \left|\frac{k\pi}{\ell_j} - \frac{n\pi}{\ell_m}\right| \cdot \left|\frac{k\pi}{\ell_j} + \frac{n\pi}{\ell_m}\right| \\
&\asymp |k(\ell_m/\ell_j) - n| \cdot k \\
&\geq \frac{c_\alpha k}{k^\alpha}
\end{aligned}$$
for every $\alpha > 1$, and the lemma follows with $\beta := \alpha - 1$.

Proof of Proposition 10.6. Using the representation (10.18) of the solutions and using the relations $\mu_{j,k} \asymp k$, the estimates (10.17) can be rewritten in the following equivalent form:

$$\sum_{j=1}^{N}\sum_{k=1}^{\infty} k^{2s}\left(|b_{j,k}|^2 + |b_{j,-k}|^2\right)$$

$$\leq c_{s,I}\int_I \left|\sum_{j=1}^{N}\sum_{k=1}^{\infty}\mu_{j,k}b_{j,k}e^{i\omega_{j,k}t} + \mu_{j,k}b_{j,-k}e^{-i\omega_{j,k}t}\right|^2 dt.$$

Let us rearrange the exponents $\pm\omega_{j,k}$ into an increasing sequence (λ_n). Excluding a set of zero measure of the N-tuples $(\ell_1, \ldots, \ell_N)$, we may assume that $\lambda_n \neq \lambda_k$ whenever $n \neq k$. Denoting the corresponding coefficients $\mu_{j,k}b_{j,\pm k}$ by b_n for brevity and using the relations $\mu_{j,k} \asymp k$, we have to establish the following estimate:

$$\sum_{n=-\infty}^{\infty} |\lambda_n|^{s-1}|b_n|^2 \leq c_{s,I}\int_I \left|\sum_{n=-\infty}^{\infty} b_n e^{i\lambda_n t}\right|^2 dt. \tag{10.22}$$

It follows from the structure of the exponents λ_n that an interval of length r contains at most $2\ell_j\sqrt{r/2}/\pi$ elements of the the family $(\pm\omega_{j,k})$, $k = 1, 2, \ldots$, the best choice for the interval being $[-r/2, r/2]$. Hence

$$n^+(r) \leq \sum_{j=1}^{N} \frac{2\sqrt{r/2}\,\ell_j}{\pi}$$

for all $r > 0$, whence

$$D^+ = \lim_{r \to \infty} \frac{n^+(r)}{r} = 0.$$

Now, given an interval I, choose γ, M as in Proposition 9.3 (p. 175) and then choose γ' such that $N\gamma' < \pi/\ell_j$ for all j. Then no chain of close exponents is longer than N; i.e., $A_j = \emptyset$ for all $j > N$. Introducing the functions $e_n(t)$ as in Theorem 9.4 (p. 177), we have the estimates

$$\int_I \left| \sum_{n=-\infty}^{\infty} a_n e_n(t) \right|^2 dt \asymp \sum_{n=-\infty}^{\infty} |a_n|^2. \tag{10.23}$$

If $m \in A_j$ for some $j = 1, \ldots, N$, then rewriting the sums according to the equation

$$\sum_{n=m}^{m+j-1} b_n e^{i\lambda_n t} = \sum_{n=m}^{m+j-1} a_n e_n(t),$$

we obtain from the structure of the divided differences the inequality

$$\sum_{n=m}^{m+j-1} |b_n|^2 \cdot \min\{|\lambda_p - \lambda_q| : m \le p < q < m+j\}^{2N-2} \le C \sum_{n=m}^{m+j-1} |a_n|^2$$

with a constant C independent of the choice of m. Furthermore, applying Lemma 10.7, we have

$$\min\{|\lambda_p - \lambda_q| : m \le p < q < m+j\}^{2N-2} \ge C_\alpha \max\{|\lambda_n| : m \le n < m+j\}^{-\beta}$$

for every $\beta > 0$. Therefore, we deduce from the last inequality that

$$\sum_{n=m}^{m+j-1} |\lambda_n|^{-\beta} |b_n|^2 \le C_\beta \sum_{n=m}^{m+j-1} |a_n|^2,$$

and then

$$\sum_{n=-\infty}^{\infty} |\lambda_n|^{-\beta} |b_n|^2 \le C_\beta \sum_{n=-\infty}^{\infty} |a_n|^2 \tag{10.24}$$

for all $\beta > 0$. Now (10.23) and (10.24) imply (10.22) for every $s < 1$.

Remark. The above proof can be adapted to a more general situation in which the system contains both strings and beams; see Sikolya [125].

10.4 Observability of Spherical Shells

The study of observability of different shell models is relatively recent; see, e.g., [39], [40], [41], [42]. In this section we present, following [102], optimal results for spherical shells with a central hole.

By the Love–Koiter linear shell theory ([105], [130]) we can formulate the mathematical model of a spherical cup of opening angle $0 < \theta_0 < \pi$ with a hole of opening angle $0 < \theta_1 < \theta_0$. In the case $\theta_0 = \frac{\pi}{2}$, a similar analysis can be done also in the absence of a hole; see [99]. We consider only axially symmetric deformations. Then the meridional and radial displacements $u(t, \theta)$ and $w(t, \theta)$ of a point P, belonging to the middle surface of the shell, satisfy in $(\theta_1, \theta_0) \times \mathbb{R}$ the following coupled system of partial differential equations:

$$\begin{cases} du_{tt} - \mathcal{L}(u) + (1 + \nu)w' - e\mathcal{L}(u + w') = 0, \\ dw_{tt} - \frac{1+\nu}{\sin\theta}(u\sin\theta)' + \frac{e}{\sin\theta}[\mathcal{L}(u + w')\sin\theta']' + 2(1 + \nu)w = 0, \end{cases} \quad (10.25)$$

where $'$ and the subscript t stand for the derivatives with respect to θ and t,

$$\mathcal{L}(v) := v'' + v'\cot\theta - (\nu + \cot^2\theta)v,$$

and d, c, ν are given constants. More precisely, denoting by R and h the radius and the half-thickness of the middle surface, by λ and η the Lamé constants, by d_0 the density and by E the Young modulus, we have

$$c = \frac{h^2}{3R^2}, \quad \nu = \frac{\lambda}{\lambda + 2\eta}, \quad \text{and} \quad d = \frac{d_0 E}{1 - \nu^2}R^2.$$

Note that $-1 < \nu < 1/2$ and $c, d > 0$.

According to the Hilbert uniqueness method, the exact controllability of this system holds in suitable function spaces, provided a special uniqueness property is satisfied. This was explained for the present context in [41], so that in this section we study only the required uniqueness of the solutions of (10.25) completed by the following boundary and initial conditions:

$$\begin{cases} u(t, \theta_0) = u(t, \theta_1) = 0, \\ w'(t, \theta_0) = w'(t, \theta_1) = 0, \\ \mathcal{L}(u + w')(t, \theta_0) = \mathcal{L}(u + w')(t, \theta_1) = 0, \end{cases} \quad t \in \mathbb{R}, \quad (10.26)$$

$$\begin{cases} u(0, \theta) = u_0, \quad u_t(0, \theta) = u_1, \\ w(0, \theta) = w_0, \quad w_t(0, \theta) = w_1, \end{cases} \quad \theta_1 < \theta < \theta_0. \quad (10.27)$$

It follows from more general results established in [42] that the problem (10.25)–(10.27) is well-posed in the Hilbert space $V \times \mathcal{H}$ defined by

$$V := H_0^1(\theta_1, \theta_0) \times (H^2 \cap H_0^1)(\theta_1, \theta_0)$$

and

$$\mathcal{H} := L^2(\theta_1, \theta_0) \times L^2(\theta_1, \theta_0).$$

In [42] more complex spaces are used, but under the present assumption $\theta_1 > 0$ they are equivalent to the above ones. Our main result is the following:

Theorem 10.8. *For all but countably many exceptional values of c, the following uniqueness property holds. If a solution of* (10.25)–(10.27) *satisfies*

$$w(t, \theta_0) = 0, \quad 0 < t < T,$$

for some $T > 2\sqrt{d}(\theta_0 - \theta_1)$, then in fact $v = (u, w)$ vanishes identically in $(\theta_1, \theta_0) \times \mathbb{R}$.

Remark. The same conclusion was obtained in [39] for the particular case of the half-sphere ($\theta_0 = \pi/2$, $\theta_1 = 0$), for some very particular choices of the parameters. The proof had two important ingredients:

- By the particular choice of the angles, the eigenfunctions of the infinitesimal generator $\mathcal{A}$ of the corresponding semigroup have an explicit representation by Legendre polynomials.
- By the choice of the parameters, the spectrum of $\mathcal{A}$ satisfies a crucial gap condition, enabling one to apply Ingham's Theorem 4.3.

In order to treat the present general case, we have to modify substantially our approach:

- Without determining explicitly the eigenfunctions and eigenvalues of $\mathcal{A}$, we can establish the existence of a Riesz basis of $V \times \mathcal{H}$, formed by eigenfunctions of $\mathcal{A}$, and we can obtain sufficiently precise information on the distribution of the corresponding eigenvalues by applying the spectral theory of ordinary differential operators as described by Titchmarsh in [131].
- Study of the eigenvalues shows that the gap condition needed for the application of Ingham's theorem is not satisfied in general. However, a weaker gap condition still holds, and this is still sufficient for our purposes because we may apply Theorem 9.4 (p. 177).

Let us clarify the structure of the solutions of (10.25)–(10.27). We refer to [123] for a study of the spectrum in the general case. In the present particular case, following [130], it is useful to introduce a primitive s of u with respect to θ and to use the differential operator

$$\mathcal{D}(s) = s'' + s' \cot \theta + 2s.$$

Then, setting also

$$k := (1 + c)(1 + \nu)$$

for brevity, (10.25) can be rewritten in a more convenient form:

$$\begin{cases} ds_{tt} = \mathcal{D}(s) + (c\mathcal{D} - k)(s + w), \\ dw_{tt} = (1 + \nu)\mathcal{D}(s) - (c\mathcal{D}^2 - c(3 + \nu)\mathcal{D} + 2k)(s + w). \end{cases} \tag{10.28}$$

Consider the following eigenvalue problem:

$$\begin{cases} -\mathcal{D}(f_j) = \alpha_j f_j & \text{in } (\theta_1, \theta_0), \\ f_j'(\theta_0) = f_j'(\theta_1) = 0. \end{cases} \tag{10.29}$$

As a consequence of our assumption $0 < \theta_1 < \theta_0 < \pi$, the coefficients of $\mathcal{D}$ are continuous on the compact interval $[\theta_1, \theta_0]$. (The assumption on the existence of a hole is crucial here.) We may therefore apply the spectral theory as developed in the first chapter of Titchmarsh's book [131]. Thus there exists a Riesz basis $f_0, f_1, \ldots$ of $L^2(\theta_1, \theta_0)$, formed by eigenfunctions of the problem (10.29). Furthermore, the following asymptotic relations are satisfied as $j \to \infty$:

$$\sqrt{\alpha_j} = \frac{j\pi}{\theta_0 - \theta_1} + O\left(\frac{1}{j}\right), \tag{10.30}$$

$$f_j = \sqrt{\frac{2}{\theta_0 - \theta_1}} \cos\left(\frac{j\pi\theta}{\theta_0 - \theta_1}\right) + O\left(\frac{1}{j}\right).$$

Rewriting (10.28) in the operational form

$$d v_{tt} = \mathcal{A}v, \quad v = (s, w),$$

and using these eigenfunctions, we can find a Riesz basis of $\mathcal{V} \times \mathcal{H}$, formed by eigenfunctions of the form $(\omega_j f_j, f_j)$ of $\mathcal{A}$. Indeed, the equation $\mathcal{A}(\omega_j f_j, f_j) = \lambda_j(\omega_j f_j, f_j)$ leads to the algebraic system

$$\begin{pmatrix} (1+c)\alpha_j + k + \lambda_j & c\alpha_j + k \\ c\alpha_j^2 + c(3+\nu)\alpha_j + (1+\nu)\alpha_j + 2k & c\alpha_j^2 + c(3+\nu)\alpha_j + 2k + \lambda_j \end{pmatrix} \begin{pmatrix} \omega_j \\ 1 \end{pmatrix} = 0.$$

Proceeding as, e.g., in [76], we have two solutions:

$$\lambda_j^\pm = \frac{1}{2}\left(-B_j \pm \sqrt{B_j^2 - 4C_j}\right)$$

with

$$B_j = c\alpha_j^2 + [(1+c) + c(3+\nu)]\alpha_j + 3(1+c)(1+\nu),$$
$$C_j = c\alpha_j^3 + 2c\alpha_j^2 + (1+c)(1-\nu^2)\alpha_j,$$

and

$$\omega_j^\pm = \frac{c\alpha_j + (1+c)(1+\nu)}{\lambda_j^\pm + (1+c)\alpha_j + (1+c)(1+\nu)}.$$

Moreover, we may assume that the numbers $\lambda_0^\pm, \lambda_1^\pm, \ldots$ are distinct and different from zero (this holds for all but countably many exceptional values of c).

Since $\alpha_j \to \infty$, one obtains easily the asymptotic relations

$$\lambda_j^+ \sim -\alpha_j, \quad \lambda_j^- \sim -c\alpha_j^2, \tag{10.31}$$

and hence

$$\omega_j^+ \sim 1, \quad \omega_j^- \sim -1/\alpha_j.$$

Applying Proposition 2.1 from [77], we conclude that the vectors

$$(\omega_j^\pm f_j, f_j), \quad j = 0, 1, \ldots,$$

form a Riesz basis in $\mathcal{H}$ and that the solutions of (10.26)–(10.28) (with $u = s'$) are given by the series

$$(s, w)(t) = \sum_j \left(a_j e^{\sqrt{\lambda_j^+/dt}} + b_j e^{-\sqrt{\lambda_j^+/dt}} \right) (\omega_j^+ f_j, f_j)$$

$$+ \sum_j \left(c_j e^{\sqrt{\lambda_j^-/dt}} + d_j e^{-\sqrt{\lambda_j^-/dt}} \right) (\omega_j^- f_j, f_j) \quad (10.32)$$

with suitable complex coefficients a_j, b_j, c_j, and d_j, depending on the initial data.

Now turning to the proof of the uniqueness theorem, we begin by formulating a special case of Theorem 9.4 (p. 177). Let $(\lambda_n)_{n=-\infty}^\infty$ be a strictly increasing sequence of real numbers. Assume that there exists a number $\gamma' > 0$ such that

$$\lambda_{n+2} - \lambda_n \geq 2\gamma'$$

for all n. Set

$$A_1 := \{n \in \mathbb{Z} \ : \ \lambda_n - \lambda_{n-1} \geq \gamma' \text{ and } \lambda_{n+1} - \lambda_n \geq \gamma'\},$$
$$A_2 := \{n \in \mathbb{Z} \ : \ \lambda_n - \lambda_{n-1} \geq \gamma' \text{ and } \lambda_{n+1} - \lambda_n < \gamma'\},$$

and consider the sums of the form

$$f(t) = \sum_n b_n e^{i\lambda_n t} \quad (10.33)$$

with complex coefficients b_n. We consider only "finite" sums; i.e., we assume that only finitely many coefficients are different from zero. Put

$$E(f) := \sum_{n \in A_1} |b_n|^2 + \sum_{n \in A_2} \left[|b_n + b_{n+1}|^2 + (\lambda_{n+1} - \lambda_n)^2 (|b_n|^2 + |b_{n+1}|^2) \right]$$

for brevity. Furthermore, set

$$D^+ := \lim_{r \to \infty} \frac{n^+(r)}{r},$$

where $n^+(r)$ denotes the largest number of terms of the sequence (λ_n) contained in an interval of length r.

The following result is a special case of Theorem 9.4 (p. 177).

Theorem 10.9. *For every bounded interval I of length $|I| > 2\pi D^+$ we have*

$$\int_I |f(t)|^2 \, dt \asymp E(f) \tag{10.34}$$

for all functions f of the form (10.33).

Remarks.

- By a standard density argument, the estimates (10.34) also remain valid for all *infinite* sums such that $E(f) < \infty$.
- Using Theorem 6.2 (p. 93), the above theorem remains valid if there is also a finite number of *nonreal* exponents λ_n.

Now we are ready to prove Theorem 10.8. Let $T > 2\sqrt{d}(\theta_0 - \theta_1)$ and assume that $w(\theta_0, t) = 0$ for all $0 < t < T$. Then, using the representation (10.32) we have

$$\sum_j a_j f_j(\theta_0) e^{\sqrt{\lambda_j^+}/dt} + b_j f_j(\theta_0) e^{-\sqrt{\lambda_j^+}/dt}$$

$$+ c_j f_j(\theta_0) e^{\sqrt{\lambda_j^-}/dt} + d_j f_j(\theta_0) e^{-\sqrt{\lambda_j^-}/dt} = 0$$

for all $0 < t < T$.

Let us arrange the numbers $\pm\sqrt{\lambda_j^\pm}$ into a new strictly increasing sequence (λ_n), and let us apply Theorem 10.9 and the above remarks. As a consequence of the asymptotic relations (10.30) and (10.31) we have $D^+ = \sqrt{d}(\theta_0 - \theta_1)/\pi$. Since $T > 2\pi D^+$, we conclude that

$$a_j f_j(\theta_0) = b_j f_j(\theta_0) = c_j f_j(\theta_0) = d_j f_j(\theta_0) = 0$$

for all j. Since the variational problem (10.29) is regular, none of the numbers $f_j(\theta_0)$ is equal to zero. Hence all coefficients a_j, b_j, c_j, and d_j vanish. Using again the representation (10.32) we conclude that the solution (s, w) and then also (u, w) vanish identically.

Remark. There exist effectively exceptional values of the parameters c. Indeed, one can find by direct computation two different indices $j < k$ and values c, ν such that $\lambda_j^+ = \lambda_k^-$. Denote this common value by λ. The formula

$$(s, w)(t) = e^{\sqrt{\lambda}/dt}\big(f_k(\theta_0)(\omega_j^+ f_j, f_j) - f_j(\theta_0)(\omega_k^- f_k, f_k)\big)$$

defines a nontrivial solution of (10.25)–(10.27) for which $w(\theta_0, t) = 0$ for all real t.

References

1. S. Agmon, *Elliptic Boundary Value Problems*, Van Nostrand, Princeton, New Jersey, 1965.
2. F. Alabau, P. Cannarsa, and V. Komornik, *Indirect internal stabilization of weakly coupled evolution equations*, J. Evol. Equ. 2 (2002), no. 2, 127–150.
3. F. Ammar Khodja and A. Benabdallah, *Sufficient conditions for uniform stabilization of second order equations by dynamical controllers*, Dynam. Contin. Discrete Impuls. Systems 7 (2000), no. 2, 207–222.
4. S.A. Avdonin and S.A. Ivanov, *Families of Exponentials*, Cambridge University Press, 1995.
5. S.A. Avdonin and W. Moran, *Simultaneous control problems for systems of elastic strings and beams*, Systems & Control Letters 44 (2) (2001), 147–155.
6. C. Baiocchi, V. Komornik, and P. Loreti, *Théorèmes du type Ingham et application à la théorie du contrôle*, C. R. Acad. Sci. Paris Sér. I Math. 326 (1998), 453–458.
7. C. Baiocchi, V. Komornik, and P. Loreti, *Ingham type theorems and applications to control theory*, Bol. Un. Mat. Ital. B (8) 2 (1999), no. 1, 33–63.
8. C. Baiocchi, V. Komornik, and P. Loreti, *Généralisation d'un théorème de Beurling et application à la théorie du contrôle*, C. R. Acad. Sci. Paris Sér. I Math. 330 (4) (2000) 281–286.
9. C. Baiocchi, V. Komornik, and P. Loreti, *Ingham-Beurling type theorems with weakened gap conditions*, Acta Math. Hungar. 97 (1-2) (2002), 55–95.
10. J. Ball and M. Slemrod, *Nonharmonic Fourier series and the stabilization of distributed semi-linear control systems*, Comm. Pure Appl. Math. 37 (1979), 555–587.
11. C. Bardos, G. Lebeau, and J. Rauch, *Sharp sufficient conditions for the observation, control and stabilization of waves from the boundary*, SIAM J. Control Optim. 30 (1992), 1024–1065.
12. A. Benabdallah and A. Soufyane, *Uniform stability and stabilization of linear thermoelastic systems*, J. Dynam. Control Systems 6 (2000), no. 4, 543–560.
13. V. Bernstein *Leçons sur les progrès récents de la théorie des séries de Dirichlet professées au Collège de France*, Gauthier-Villars, Paris, 1933.
14. J.N.J.W.L. Carleson and P. Malliavin, editors, *The Collected Works of Arne Beurling*, Volume 2, Birkhäuser, 1989.

15. R. Bey, J.-P. Lohéac, and M. Moussaoui, *Singularities of the solution of a mixed problem for a general second order elliptic equation and boundary stabilization of the wave equation*, J. Math. Pures Appl. (9) 78 (1999), 10, 1043–1067.

16. F. Bourquin, private communication, 1996.

17. F. Bourquin, J.-S. Briffaut, and M. Collet, *On the feedback stabilization: Komornik's method*, Proceedings of the Second International Conference on Active Control in Mechanical Enginering, Lyon, 22–23 October 1997.

18. F. Bourquin, J.-S. Briffaut, and J. Urquiza, *Contrôlabilité exacte et stabilisation rapide des structures : aspects numériques*, Actes de l'Ecole CEA INRIA EDF sur les matériaux intelligents, April 1997.

19. F. Bowman, *Introduction to Bessel Functions*, Dover, New York, 1958.

20. H. Brezis, *Analyse fonctionnelle. Théorie et applications*, Masson, Paris, 1983.

21. J.-S. Briffaut, *Méthodes numériques pour le contrôle et la stabilisation rapide des grandes structures flexibles*, Thèse de doctorat de l'École Nationale des Ponts et Chaussées, Paris, June 14, 1999.

22. J.W.S. Cassels, *An Introduction to Diophantine Approximation*, Cambridge Tracts in Mathematics and Mathematical Physics, No. 45. Cambridge University Press, New York, 1957.

23. C. Castro and E. Zuazua, *Une remarque sur les séries de Fourier non-harmoniques et son application à la contrôlabilité des cordes avec densité singulière*, C. R. Acad. Sci. Paris Sér. I 322 (1996), 365–370.

24. M. Cherkaoui, F. Conrad, and N. Yebari, *Points d'équilibre pour une équation des ondes avec contrôle frontière contenant un terme intégral*, Port. Math. (N.S.) 59 (2002), no. 3, 351–370.

25. P.-G. Ciarlet, *Mathematical Elasticity*, Vol. 2, North Holland, Amsterdam, 1988.

26. E.A. Coddington and N. Levinson, *Theory of Ordinary Differential Equations*, McGraw-Hill, New York, 1955.

27. R. Courant and D. Hilbert, *Methods of Mathematical Physics I*, John Wiley & Sons, New York, 1989.

28. R. Dáger and E. Zuazua, *Controllability of star-shaped networks of strings*, C. R. Acad. Sci. Paris Sér. I Math. 332 (2001), 621–626.

29. R. Dáger and E. Zuazua, *Controllability of tree-shaped networks of strings*, C. R. Acad. Sci. Paris Sér. I 332 (2001), 1087–1092.

30. R. Dáger and E. Zuazua, *Spectral boundary controllability of networks of strings*, C. R. Acad. Sci. Paris Sér. I 334 (2002), 545–550.

31. R. Dautray and J.-L. Lions, *Mathematical Analysis and Numerical Methods for Science and Technology*, Vol. 1–6, Springer-Verlag, Berlin, 1990–93.

32. S. Dolecki and D.L. Russell, *A general theory of observation and control*, SIAM J. Control Opt. 15 (1977), 185–220.

33. K.J. Falconer, *The Geometry of Fractal Sets*, Cambridge University Press, 1986.

34. H.O. Fattorini and D.L. Russell, *Uniform bounds on biorthogonal functions for real exponentials with an application to the control theory of parabolic equations*, Quart. Appl. Math. 32 (1974), 45–69.

35. F. Flandoli, *A new approach to the L-Q-R problem for hyperbolic dynamics with boundary control*, Lect. Notes Control Information Sciences 102, Springer-Verlag, Berlin, Heidelberg, New York, 1987, 89-111.

36. A.V. Fursikov and O.Yu. Imanuvilov, *Controllability of Evolution Equations*, Lecture Notes of the Research Institute of Mathematics, Global Analysis Research Center, Vol. 34, Seoul National University, Korea, 1996.

37. N. Garofalo and F. Lin, *Monotonicity properties of variational integrals, A_p weights and unique continuation*, Indiana Univ. Math. J. 35 (1986), 245–268.

38. N. Garofalo and F. Lin, *Unique continuation for elliptic operators: a geometric-variational approach*, Comm. Pure Appl. Math. 40 (1987), 347–366.

39. G. Geymonat, P. Loreti, and V. Valente, *Introduzione alla controllabilità esatta per la calotta sferica*, Quaderno IAC 8/1989.

40. G. Geymonat, P. Loreti, and V. Valente, *Exact Controllability of a shallow shell model*, Int. Series of Num. Math. 107 (1992), 85–97.

41. G. Geymonat, P. Loreti, and V. Valente, *Exact controllability of a spherical shell via harmonic analysis*, in "Boundary Values Problems for Partial Differential Equations and Applications," J.-L. Lions, C. Baiocchi, editors, Masson, Paris, 1993.

42. G. Geymonat, P. Loreti, and V. Valente, *Spectral problems for thin shells and exact controllability*, in Spectral Analysis of Complex Structures, Collection "Travaux en cours," Hermann, Paris, 49 (1995), 35–57.

43. G. Geymonat and V. Valente, *A noncontrollability result for systems of mixed order*, SIAM J. Control Control Optim. 39 (2000), no. 3, 661–672.

44. K.D. Graham and D.L. Russell, *Boundary value control of the wave equation in a spherical region*, SIAM J. Control 13 (1975), 174–196.

45. P. Grisvard, *Contrôlabilité exacte des solutions de l'équation des ondes en présence de singularités*, J. Math. Pures Appl. (9) 68 (1989), 215–259.

46. P. Halmos, *Introduction to Hilbert Space and the Theory of Spectral Multiplicity*, Chelsea, New York, 1957.

47. S. Hansen, *Bounds on functions biorthogonal to sets of complex potentials; control of damped elastic systems*, J. Math. Anal. Appl. 158 (1991), 487–508.

48. A. Haraux, *Séries lacunaires et contrôle semi-interne des vibrations d'une plaque rectangulaire*, J. Math. Pures Appl. 68 (1989), 457–465.

49. A. Haraux, *Quelques méthodes et résultats récents en théorie de la contrôlabilité exacte*, Rapport de recherche No. 1317, INRIA Rocquencourt, Octobre 1990.

50. L.F. Ho, *Observabilité frontière de l'équation des ondes*, C. R. Acad. Sci. Paris Sér. I Math. 302 (1986), 443–446.

51. M. Horváth, *Vibrating strings with free ends*, Acta Math. Acad. Sci. Hungar. 51 (1988), 171–198.

52. A. E. Ingham, *Some trigonometrical inequalities with applications in the theory of series*, Math. Z. 41 (1936), 367–379.

53. E. Isaacson and H.B. Keller, *Analysis of Numerical Methods*, John Wiley & Sons, New York, 1966.

54. S. Jaffard, *Contrôle interne exact des vibrations d'une plaque carrée*, C. R. Acad. Sci. Paris Sér. I Math. 307 (1988), 759–762.

55. S. Jaffard, *Contrôle interne exact des vibrations d'une plaque rectangulaire*, Portugalia Math. 47 (1990), 423–429.

56. S. Jaffard, M. Tucsnak, and E. Zuazua, *On a theorem of Ingham*, J. Fourier Anal. Appl. 3 (1997), no. 5, 577–582.

57. S. Jaffard, M. Tucsnak, and E. Zuazua, *Singular internal stabilization of the wave equation*, J. Differential Equations 145 (1998), 1, 184–215.

58. V. Jarník, *Über die simultanen diophantischen Approximationen*, Math. Z. 33 (1931), 505–543.

59. I. Joó, *Contrôlabilité exacte et propriétés d'oscillation de l'équation des ondes par analyse non harmonique*, C. R. Acad. Sci. Paris Sér. I Math. 312 (1991), 119–122.

60. I. Joó, *On the control of a circular membrane and related problems*, Annales Univ. Sci. Budapest. Sect. Math. 34 (1991), 231–266.

61. J.-P. Kahane, *Pseudo-périodicité et séries de Fourier lacunaires*, Ann. Sci. de l'E.N.S. 79 (1962), 93–150.

62. V. Komornik, *Contrôlabilité exacte en un temps minimal*, C. R. Acad. Sci. Paris Sér. I Math. 304 (1987), 223–225.

63. V. Komornik, *A new method of exact controllability in short time and applications*, Ann. Fac. Sci. Toulouse 10 (1989), 415–464.

64. V. Komornik, *On the exact internal controllability of a Petrowsky system*, J. Math. Pures Appl. (9) 71 (1992), 331–342.

65. V. Komornik, *On the zeros of Bessel type functions and applications to exact controllability problems*, Asymptotic Anal. 5 (1991), 115–128.

66. V. Komornik, *A generalization of Ingham's inequality*, Differential equations and its applications (Budapest, 1991), Colloq. Math. Soc. János Bolyai 62 (1991), 213–217.

67. V. Komornik, *Exact Controllability and Stabilization. The Multiplier Method*, Masson, Paris, and John Wiley & Sons, Chicester, 1994.

68. V. Komornik, *Stabilisation frontière rapide de systèmes distribués linéaires*, C. R. Acad. Sci. Paris Sér. I Math. 321 (1995), 433–437.

69. V. Komornik, *Stabilisation rapide de problèmes d'évolution linéaires*, C. R. Acad. Sci. Paris Sér. I Math. 321 (1995), 581–586.

70. V. Komornik, *Rapid boundary stabilization of linear distributed systems*, SIAM J. Control Optim. 35 (1997), 1591–1613.

71. V. Komornik, *Rapid boundary stabilization of Maxwell's equations*, Équations aux dérivées partielles et applications, Articles dédiés à Jacques-Louis Lions, Gauthier-Villars, Paris, 1998, 611–622.

72. V. Komornik, *Boundary observability, controllability and stabilizability of linear distributed systems with a time reversible dynamics*, Control Cybernet. 28 (1999), 4, 813–838.

73. V. Komornik, *Control and stabilization of coupled linear systems*, Optimal Control and Partial Differential Equations. In honour of Professor Alain Bensoussan's 60th birthday. J. L. Menaldi et al. (eds.), IOS Press, Amsterdam, 2000, 374–383.

74. V. Komornik and P. Loreti, *Observabilité frontière de systèmes couplés par analyse non harmonique vectorielle*, C. R. Acad. Sci. Paris Sér. I Math. 324 (1997), 895–900.

75. V. Komornik and P. Loreti, *Partial observability of coupled linear systems*, Acta Math. Hungar. 86 (2000), 49–74.

76. V. Komornik and P. Loreti, *Ingham type theorems for vector-valued functions and observability of coupled linear systems*, SIAM J. Control Optim. 37 (1998), 461–485.

77. V. Komornik and P. Loreti, *Observability of compactly perturbed systems*, J. Math. Anal. Appl. 243 (2000), 409–428.

78. V. Komornik and P. Loreti, *A constructive approach for the observability of coupled linear systems*, Control and partial differential equations (Marseille–Luminy, 1997), ESAIM: Proceedings 4 (1998), 171–179.

79. V. Komornik and P. Loreti, *Dirichlet series and observability problems*, Systems & Control Letters 48 (2003), 3–4, 221–227.
80. V. Komornik and P. Loreti, *Boundary observability of compactly perturbed systems*, Proceedings of the 8th Conference on Control of Distributed Parameter Systems (Graz, 2001), Internat. Ser. Numer. Math. (143) 2002, 219–230.
81. V. Komornik, P. Loreti, and G. Vergara Caffarelli, *A comparative study of the control of two beam models*, Bol. Soc. Paran. Mat. (3s.) 20 (2002), 1/2, 59–72.
82. V. Komornik, P. Loreti, and E. Zuazua, *On the control of coupled linear systems*, Control and estimation of distributed parameter systems (Vorau, 1996), Internat. Ser. Numer. Math. 126 (1998), 183–189.
83. V. Komornik and E. Zuazua, *A direct method for the boundary stabilization of the wave equation*, J. Math. Pures Appl. (9) 69 (1990), 33–54.
84. W. Krabs, *On Moment Theory and Controllability of One-Dimensional Vibrating Systems and Heating Processes*, Lect. Notes in Control and Inform. Sci., vol. 173, Springer-Verlag, Berlin, 1992.
85. W. Krabs, G. Leugering, and T.I. Seidman, *On boundary controllability of a vibrating plate*, Appl. Math. Optim. 13 (1985), 205–229.
86. J.E. Lagnese, *Boundary Stabilization of Thin Plates*, SIAM Studies in Appl. Math., Philadelphia, 1989.
87. J.L. Lagnese and G. Leugering, *Uniform stabilization of a nonlinear beam by nonlinear boundary feedback*, J. Diff. Equations 91 (1991), 355–388.
88. J.E. Lagnese and J.-L. Lions, *Modelling, Analysis and Control of Thin Plates*, Masson, Paris, 1988.
89. H.J. Landau, *Necessary density conditions for sampling and interpolation of certain entire functions*, Acta Math. 117 (1967), 37–52.
90. I. Lasiecka and R. Triggiani, *Regularity of hyperbolic equations under $L_2(0, T; L_2(\Gamma))$ boundary terms*, Appl. Math. and Optimiz. 10 (1983), 275–286.
91. I. Lasiecka and R. Triggiani, *Control theory for partial differential equations: continuous and approximation theories I-II*, Encyclopedia of Mathematics and Its Applications, 74–75. Cambridge University Press, 2000.
92. G. Lebeau, *Contrôle de l'équation de Schrödinger*, J. Math. Pures Appl. 71 (1992), 267–291.
93. Ta-tsien Li and B. Rao, *Local exact boundary controllability for a class of quasilinear hyperbolic systems*, Dedicated to the memory of Jacques-Louis Lions. Chinese Ann. Math. Ser. B 23 (2002), no. 2, 209–218.
94. J.-L. Lions, *Contrôle des systèmes distribués singuliers*, Gauthier-Villars, Paris, 1983.
95. J.-L. Lions, *Contrôlabilité exacte des systèmes distribués*, C. R. Acad. Sci. Paris Sér. I Math. 302 (1986), 471–475.
96. J.-L. Lions, *Exact controllability, stabilizability, and perturbations for distributed systems*, Siam Rev. 30 (1988), 1–68.
97. J.-L. Lions, *Contrôlabilité exacte et stabilisation de systèmes distribués I-II*, Masson, Paris, 1988.
98. J.-L. Lions and E. Magenes, *Non-homogeneous Boundary Value Problems and Applications I-III*, Die Grundlehren der mathematischen Wissenschaften, Bände 181–183. Springer-Verlag, New York-Heidelberg, 1972- 73.
99. P. Loreti, *Application of a new Ingham type theorem to spherical shells*, Proceedings of the 11th IFAC International Workshop *Control Applications of Optimization* (St. Petersburg, Russia, July 3–6, 2000), Pergamon, 2000.

100. P. Loreti, *Partial rapid stabilization of linear distributed systems*, Annales Univ. Sci. Budapest. 42 (1999), 93–100.

101. P. Loreti, *Some applications of new Ingham type theorems to control theory*, Panamerican Math. J. 11 (2001), 2, 95–104.

102. P. Loreti, *Exact controllability of shells in minimal time*, Rend. Mat. Acc. Lincei (9) 12 (2001), 43–48.

103. P. Loreti, *On some gap theorems*, submitted to the Proceedings of the 11th General Meeting of the European Women in Mathematics (EWM), CIRM (Centre International de Rencontres Mathématiques), November 3–7, 2003.

104. P. Loreti and V. Valente, *Partial exact controllability for spherical membranes*, SIAM J. Control Optim. 35 (1997), 641–653.

105. A.E.H. Love, *A Treatise on the Mathematical Theory of Elasticity*, Dover, New York, 1944.

106. J. Marczinkiewicz and A. Zygmund, *Proof of a gap theorem*, Duke Math. J. 4 (1938), 469–472.

107. M. Mehrenberger, *Observability of coupled linear systems*, Acta Math. Hungar. 103 (4) 2004, 321–348.

108. M. Mehrenberger, *Critical length for a Beurling type theorem*, Bol. Un. Mat. Ital. B (8), to appear.

109. N.K. Nikolskii, *A Treatise on the Shift Operator*, Springer, Berlin, 1986.

110. E. Oudet, *Quelques résultats en optimisation de forme et stabilisation*, Thèse de doctorat de l'Université Louis Pasteur, IRMA, No. 2002–036, Strasbourg, 2002.

111. R.E.A.C. Paley and N. Wiener, *Fourier Transforms in the Complex Domain*, Amer. Math. Soc. Colloq. Publ., Vol. 19, Amer. Math. Soc., New York, 1934.

112. A. Pazy, *Semigroups of Linear Operators and Aplications to Partial Differential Equations*, Springer, New York, 1983.

113. G. Pólya, *Über die Existenz unendlich vieler singulärer Punkte auf der Konvergenzgerade gewisser Dirichletscher Reihen*, Berl. Sitz., 1923.

114. G. Pólya and G. Szegő, *Problems and Theorems in Analysis I-II*, Springer, Berlin, 1972.

115. L.S. Pontryagin, *Ordinary Differential Equations*, Addison-Wesley International Series in Mathematics, Addison-Wesley Publishing Co., Inc., Reading, Mass.-Palo Alto, Calif.-London 1962.

116. L. Ratier, *Stabilisation rapide des structures par contrôle actif, mise en oeuvre expérimentale*, Thèse de doctorat de l'École Nationale Supérieure de Cachan, Paris, December 7, 2000.

117. J. Rauch and M.E. Taylor, *Exponential decay of solutions to hyperbolic equations in bounded domains*, Indiana Univ. Math. J. 24 (1974), 79–86.

118. P.A. Raviart and J.M. Thomas, *Introduction à l'analyse numérique des équations aux dérivées partielles*, Masson, Paris, 1983.

119. F. Rellich, *Darstellung der Eigenwerte von $\Delta u + \lambda u$ durch ein Randintegral*, Math. Z. 18 (1940), 635–636.

120. F. Riesz and B. Sz.-Nagy, *Functional analysis*, Dover, New York, 1990.

121. W. Rudin, *Functional Analysis*, McGraw Hill, 1973.

122. D.L. Russell, *Controllability and stabilizability theory for linear partial differential equations. Recent progress and open questions*, SIAM Rev. 20 (1978), 639–739.

123. E. Sanchez-Palencia, *Asymptotic and spectral properties of a class of singular-stiff problems*, J. Math. Pures Appl. 71, 1992, 379–406.

124. K. Seip, *On the connection between exponential bases and certain related sequences in $l_2(-\pi, \pi)$*, J. Funct. Anal. 130 (1995), 131–160.

125. E. Sikolya, *Simultaneous observability of networks of beams and strings*, Bol. Soc. Paran. Mat. (3) 21 (2003), 1/2, 1-11.

126. S.L. Sobolev, *Partial Differential Equations of Mathematical Physics*, Dover, New York, 1989.

127. E.M. Stein and G. Weiss, *Introduction to Fourier Analysis on Euclidean Spaces*, Princeton Mathematical Series, No. 32. Princeton University Press, Princeton, N.J., 1971.

128. B. Sz.-Nagy and C. Foias, *Harmonic Analysis of Operators on Hilbert Space*, North-Holland Publishing Co., Amsterdam-London; American Elsevier Publishing Co., Inc., New York; Akadémiai Kiadó, Budapest 1970.

129. A.N. Tikhonov and A.A. Samarskii, *Equations of Mathematical Physics*, Dover, New York, 1990.

130. S. Timoshenko, *Theory of elastic stability*, McGraw-Hill, New York, 1936.

131. E.C. Titchmarsh, *Eigenfunction Expansions Associated with Second-Order Differential Equations*, Clarendon Press, Oxford, 1962.

132. P. Turán, *On an inequality*, Ann. Univ. Sci. Budapest. Eötvös Sect. Math. 1 (1959), 3–6.

133. D. Ullrich, *Divided differences and systems of nonharmonic Fourier series*, Proc. Amer. Math. Soc. 80 (1980), 47–57.

134. J. M. Urquiza, *Contrôle d'équations des ondes linéaires et quasilinéaires*, Thèse de doctorat de l'Université Paris VI, Paris, November 2, 2000.

135. G.N. Watson, *A Treatise on the Theory of Bessel Functions*, Cambridge University Press, 1962.

136. N. Wiener, *A class of gap theorems*, Ann. Scuola Norm. Sup. Pisa (2), 3 (1934), 367–372.

137. K. Yosida, *Functional Analysis*, Springer, Berlin, 1980.

138. R.M. Young, *An Introduction to Nonharmonic Fourier Series*, Academic Press, 1980.

139. E. Zuazua, *Contrôlabilité exacte en un temps arbitrairement petit de quelques modèles de plaques*, Appendice 1 in [97].

140. A. Zygmund, *Trigonometric Series I-II*, Cambridge University Press, 1959.

Index